Placer Deposits of Alaska

By EDWARD H. COBB

GEOLOGICAL SURVEY BULLETIN 1374

*An inventory of the placer mines
and prospects of Alaska, their
history and geologic setting*

CONTENTS

ILLUSTRATIONS

PLACER DEPOSITS OF ALASKA

By Edward H. Cobb

INTRODUCTION

Placer deposits, in addition to their intrinsic value, serve as indicators of areas of potential development of lode deposits. Any possibility that Alaska may again become an important source of metallic mineral commodities depends in part on an inventory of placer deposits and a knowledge of the geology of their source areas.

Knowledge of Alaska's placer deposits is far from complete. Many mining camps have been, at best, the subject of only cursory examination by geologists or mining engineers who are at liberty to publish the results of their studies. Many placers are vaguely known from unconfirmed reports that someone was working on a creek, and in many instances, a creek bearing that name cannot now be identified with any particular stream. Such data are, of course, practically valueless and were generally ignored in preparing this report. Many prospectors are reluctant to divulge data on their claims and this has undoubtedly caused some deposits to be "lost" or overlooked. Other major gaps in our knowledge result from the lack of bedrock exposures in many areas where valuable placer deposits were mined. In these areas, the determination of lode sources and their mode of occurrence during reconnaissance studies of mining districts was impossible.

The first in a series of reports published by the U.S. Geological Survey on the metallic minerals resources of Alaska was an index of the State's mineral deposits (Cobb and Kachadoorian, 1961) compiled from reports of Federal and State agencies published before 1960 that contain data on mines, prospects, and reported occurrences of metallic and nonmetallic deposits in Alaska. The Geological Survey has since published eight mineral commodity maps (U.S. Geological Survey mineral investigations resource maps) (Cobb 1960a–1960d, 1962, 1964a–1964c) and has released in open file 56 metallic mineral resources maps covering 81 of the 153 quadrangles into which Alaska has been divided for topographic mapping at a scale of 1:250,000 (1 in.=approx. 4 miles)

1

(Clark and Cobb, 1970; Cobb, 1967a–1967n, 1968a–1968v, 1969a–1969g, 1970a, 1970b; Cobb and Condon, 1970; Cobb and Matson, 1969; Cobb and Richter, 1967; Cobb and Sainsbury, 1968; Detterman and Cobb, 1969; Hoare and Cobb, 1970; MacKevett and Cobb, 1969; Matson, 1969a–1969c). The index and earlier reports are essentially specialized reference lists.

As many of the basic works were out of print and were unavailable to persons who do not have access to a large university or government library, data on metallic lodes were summarized in a report by Berg and Cobb (1967) that made use of all reports published before September 1, 1965. This report summarizes the geology and history of Alaska's placer deposits at nearly 1,000 localities and is a summary of reports available before January 1, 1970. More than 500 deposits are individually described or referred to. The lists of occurrences that accompany the 52 locality maps constitute an inventory of the State's known placers, and the references cited in those lists are a selected bibliography of the geology and history of placer developments in Alaska.

Material used in addition to reports published or placed in open file by Federal and State agencies includes a few pertinent journal articles, unpublished data of the Geological Survey, and papers that are in preliminary form but are not yet published. Most of the papers in "References Cited" that bear a 1970 or later date are in the last category.

Many of my colleagues in the Geological Survey have made major contributions through informal discussions during the past 10 years. Of particular assistance were C. L. Sainsbury and D. M. Hopkins, who contributed information on the Seward Peninsula; W. P. Brosgé, on northern Alaska and the south flank of the Brooks Range; J. M. Hoare, on southwestern Alaska; W. W. Patton, Jr., on parts of western Alaska; and B. L. Reed, on the southern Alaska Range.

ORGANIZATION AND METHOD OF PRESENTATION

This report is organized by mining regions and mining districts (pl. 1) as defined by the U.S. Bureau of Mines (Ransome and Kerns, 1954). Boundaries generally follow major drainage divides or major rivers. The Bureau of Mines classification, which is not based in any way on geology, was adopted because it is the classification used in the companion report on metallic lode deposits (Berg and Cobb, 1967) and in most statistical presentations of data on mineral production. Historically, regions and districts have undergone name and boundary changes, but in general these changes have been minor and are fairly well documented.

The 14 mining regions and 67 districts included are discussed in alphabetical order. Maps for regions and districts show all known placer deposits that can be located accurately enough to plot at the map scales used. Each locality or group of localities has a number keyed to a list giving the name of the deposits and the principal reference (or references) for each deposit. Some adjoining districts have been combined on single maps to conserve space. Where the concentration of deposits in an area is such that the occurrences cannot be shown adequately on regional or district maps, the placers are delineated on larger scale maps whose boundaries are shown on the general maps.

The geographic features mentioned in the text are shown on the maps where possible; all features appear on topographic maps published by the Geological Survey at 1:250,000 or larger scale. (Indices to these maps and copies of the topographic sheets may be obtained from the Geological Survey in Fairbanks, Alaska, Denver, Colo., or Washington, D.C.)

Descriptions of the placer deposits of each region (or each district in the case of the Yukon River region) are prefaced by brief summaries of the physiography and general geology of the area—these are supplemented by maps (reprinted from Berg and Cobb, 1967, pl. 1) showing the physiographic provinces of Alaska (based on Wahrhaftig, 1965) and distribution of the major rock units—and by brief summaries of the lode resources and history of placer mining.

Data on glaciation are mainly from a glacial map of Alaska prepared by a committee of the Geological Survey (Coulter and others, 1965); data on the extent and character of permafrost (perennially frozen ground) are from a permafrost map of Alaska by Ferrians (1965).

Data on production are scattered through many reports of the Geological Survey (particularly Smith, 1933c), the U.S. Bureau of Mines "Minerals Yearbook" (issued annually), and the Alaska Division of Mines and Geology and its predecessor State and Territorial agencies (annual and biennial reports). All of these sources and materials in the files of the Geological Survey were used in compiling production figures. Most data originally given as dollar values were converted to fine ounces of gold on the basis of gold values of $20.67 per fine ounce before 1934 and $35.00 per fine ounce thereafter. No production data for regions or districts are available for years after 1961, and data for many earlier years are fragmentary; the figures given in this report represent minima to which I have added my best estimates on the distribution of production not officially assigned to specific regions or districts.

The total production of gold from placer deposits of Alaska (1880 through 1968, the last year for which data are available) was about 20,848,000 fine ounces.

Discussion of the fineness of gold from Alaskan placers has been avoided in this report, as very few data could be added to those given by Smith (1941a) in an exhaustive summary prepared shortly before World War II.

DEFINITIONS

Although I have avoided jargon as much as possible, a few terms peculiar to the placer mining industry are used, and brief definitions of these terms follow:

Right-limit tributary, left-limit tributary. A small stream that enters a larger stream from, respectively, the right or left as seen facing down the larger stream.

Dredge. A self-contained placer-mining device that floats on a pond that moves with the dredge by excavation ahead of the dredge and filling with waste material behind it.

Hydraulic mining (hydraulicking). Excavation and movement of valuable gravel to sluiceboxes by high-pressure water jets.

Nonfloat mining. Excavation and movement of valuable gravel to sluiceboxes or other gold-separating devices by dragline, mechanically powered scraper, bulldozer, or other earth-moving machinery.

Drift mining. Excavation of valuable gravel through horizontal tunnels (drifts) driven from a shaft or from a sloping bank or wall of ravine; generally practiced in perennially frozen ground.

Hand mining. Excavation and movement of valuable gravel to a sluicebox or other simple gold-separating device by picks, hand shovels, and, in some operations, wheelbarrows.

Sluicebox. A sloping elongate box, commonly about 12 feet long, with riffles (blocks, rails, or other obstructions) in its bottom that trap the heavy minerals in gravel washed through the box by water.

Rocker. A simple gold-recovering device that is manually rocked back and forth, agitating gravel that has been shovelled into it and causing gold and other heavy minerals to be concentrated in riffles, matting, or other material in the bottom; portable and requires little water, which may be saved and reused.

Groundsluicing. Excavation mainly by running water not under high pressure.

False bedrock. A bed of impervious material, commonly clay, in stream deposits; gold may be concentrated on such a bed and the material beneath it may be barren.

ALASKA PENINSULA REGION

The Alaskan Peninsula region (pl. 1, fig. 1) is the area drained by the Ugashik River, Dago Creek, and all streams flowing into the Pacific Ocean south of Cape Kekurnoi and includes Unimak and Chirikof Islands and other islands southeast of the mainland.

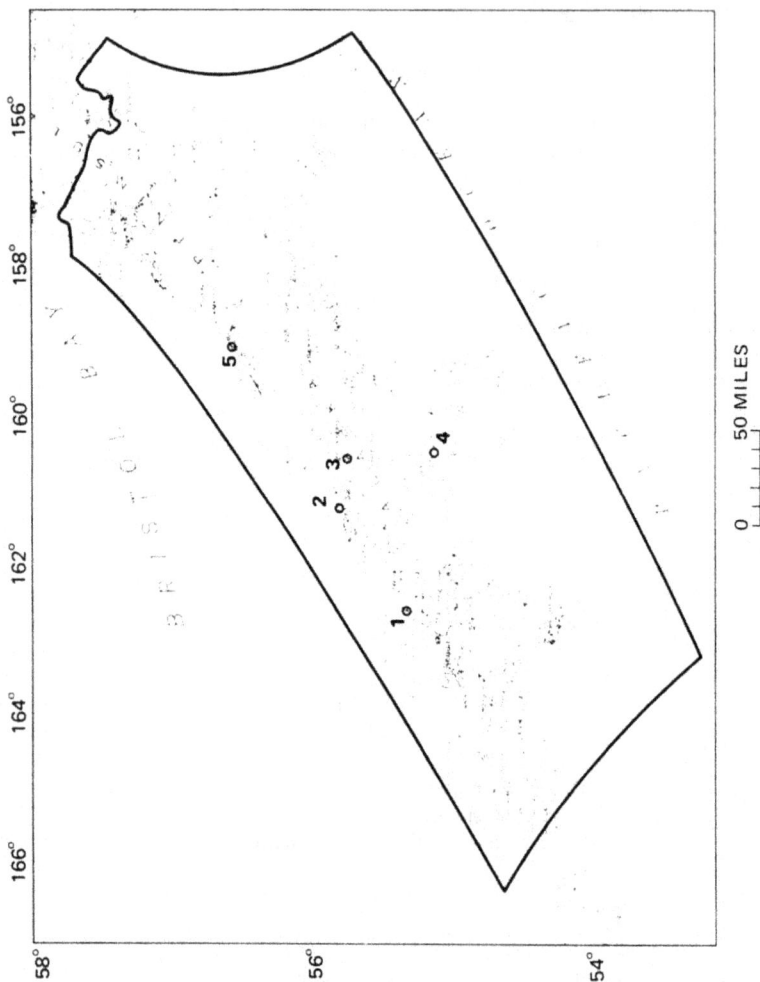

FIGURE 1.—Placer deposits in the Alaska Peninsula region.

1. Moffett Point: Berryhill (1963, p. 45–48).   4. Popof Island: Atwood (1911, p. 125).
2. Nelson Lagoon: Berryhill (1963, p. 42–46).   5. Port Heiden: Berryhill (1963, p. 33–36).
3. Port Moller: Berryhill (1963, p. 39–42).

The region is classified as a single mining district.

The region is dominated by the Aleutian Range, a series of northeast-trending ridges 1,000–4,000 feet in altitude surmounted locally by volcanoes up to 9,372 feet high. Northwestward, the range merges with a low sand- and gravel-mantled plain that has local relief of 50–250 feet.

Geologically, the Alaska Peninsula region consists of two main belts that extend for most of its length. The northwestern belt is as much as 35 miles wide near the Ugashik Lakes but is absent on the western end of Unimak Island; it is predominantly uncon-solidated Quaternary silt, sand, and gravel. The southeastern belt is made up mainly of Mesozoic and Cenozoic sedimentary and volcanic rocks and Tertiary granitic plutons (Burk, 1965). A single exposure of Permian sedimentary rocks is near Puale Bay (Hanson, 1957) ; some of the volcanic rocks also may be of Per-mian age. The entire Alaska Peninsula was glaciated during the Pleistocene Epoch but is now ice free except for some of the high-est peaks. The region is generally free of permafrost. Several of the volcanoes that surmount the Aleutian Range have been active within the past few years.

Lodes in the Alaska Peninsula region (Berg and Cobb, 1967, p. 5–7, fig. 1; Cobb, 1970b) contain gold, silver, copper, lead, and zinc. Only those on Unga Island have been worked commercially; ore worth about $2 million, chiefly in gold and silver from the Apollo mine, was produced between 1891 and 1904.

The only placer deposit in the Alaska Peninsula region for which there is a production record is an auriferous beach on Popof Island (4, fig. 1), where about 580 ounces of gold was taken out with rockers in 1904 and 1905 from a belt about three-quarters of a mile long. All gold recovered was below midtide level and most was found around large boulders near the low-tide line. Small-scale mining was reported in each of several years before World War I, but there is no record of more recent activity. The source of the gold probably is nearby lodes in intensely altered andesite. Brooks (1912, p. 37) reported beach mining on Unga Island in 1911 but did not identify where on the island or give any idea of the success of the venture.

Titaniferous magnetite and ilmenite are widespread in beach sands along the shores of Bristol Bay (Berryhill, 1963). Berryhill collected samples containing as much as 100 pounds of iron per cubic yard (calculated as content of material in place) from Mof-fett Point (1, fig. 1), Nelson Lagoon (2), Port Moller (3), and Port Heiden (5). The titania (TiO_2) content was generally less than 25 pounds per cubic yard. A few samples contained traces of fine gold. The heavy minerals in the beaches are concentrated from deposits (largely glacial) of the adjacent coastal plain.

ALEUTIAN ISLANDS REGION

The Aleutian Islands region (pl. 1, fig. 2) includes the islands west of Unimak and is considered as one district. It consists of an archipelago surmounting a submarine ridge 1,400 miles long and 20 to 60 miles wide that rises 12,000 feet above the sea floor. An arcuate line of about 40 volcanoes, 21 of which have been active in historic time (Coats, 1950), rises as much as 6,975 feet above sea level.

The region is underlain by Cenozoic basalt and andesite lava flows, tuffs, and clastic rocks and by mafic to felsic dikes, sills, and stocks. Recent work by Scholl and others (1969) has shown that rocks formerly thought to be Paleozoic(?) in age (Coats, 1956, p. 48–49) are Eocene. This region is geologically the youngest in Alaska, as all others are at least partly underlain by rocks as old as Mesozoic.

The only lodes known in the Aleutian Islands region are gold deposits (some of which may have been productive) and occurrences of metallic sulfide minerals on Unalaska and neighboring Amaknak and Sedanka Islands (Berg and Cobb, 1967, p. 7–8, fig. 2). Capps (1934, p. 149) reported rumors of copper on Salt Island, off the north shore of Atka, and speculated that the copper might have been derived from an amygdaloidal lava flow.

The only placer occurrence reported in the Aleutian Islands is a few grains of gold in tributaries of the Makushin River (fig. 2) on Unalaska Island. The occurrence is near a conspicuous gossan of severely altered and pyritized volcanic rocks near a small body of granodiorite.

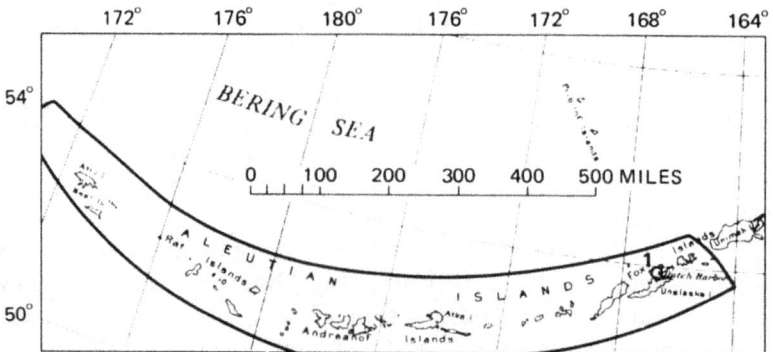

FIGURE 2.—Location of Makushin River placer occurrence (1, Drewes and others (1961, p. 657)) in the Aleutian Islands region.

BERING SEA REGION

The Bering Sea region (pl. 1, fig. 3) includes St. Lawrence, St. Matthew, and the Pribilof Islands and nearby smaller islands and offshore rocks. It is considered as one district.

The islands of the Bering Sea are mainly rolling uplands and emerged marine platforms, generally within a few hundred feet of sea level. Isolated mountain masses rise to altitudes between 800 and 900 feet, or about 2,000 feet above the shallow Bering

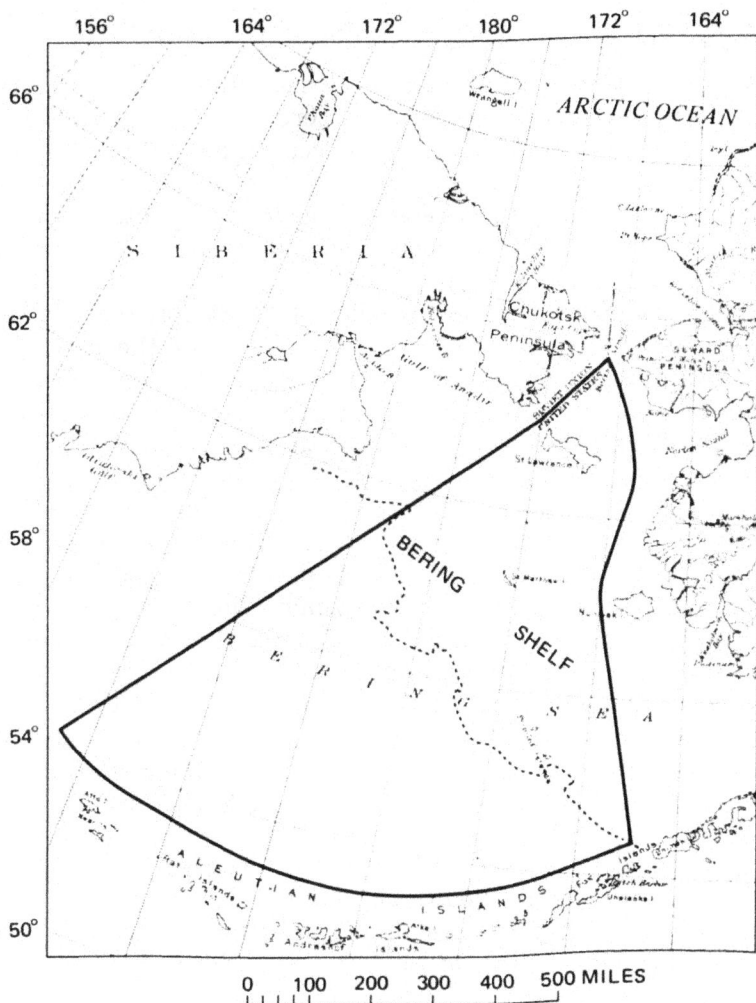

FIGURE 3.—Bering Sea region. Edge of Bering Shelf (dotted line) from Scholl and Hopkins (1969, fig. 1).

Shelf, much of which was emergent at times during the Cenozoic era. (See Hopkins (1967).)

The Pribilof Islands and St. Matthew are composed mainly of Cenozoic volcanic rocks and surficial deposits; peridotite older than the volcanic rocks underlies a small area on St. George Island in the Pribilofs (Barth, 1956; Cobb and others, 1968, p. K3–K5). St. Lawrence Island is made up of a thick section of Paleozoic and Mesozoic carbonate and clastic rocks generally similar to coeval rocks exposed in the Brooks Range of northern Alaska and in the Chukotsk Peninsula of Siberia (Patton and Dutro, 1969; Patton and Csejtey, 1970, 1971). The western part of the island contains small areas of coal-bearing Tertiary continental deposits and Cretaceous and Tertiary volcanic rocks. Tertiary(?) and Quaternary basaltic rocks cover the older rocks in central St. Lawrence Island. Permian gabbro and diabase, hypabyssal phases related to some of the volcanic rocks, and Cretaceous monzonitic plutons invaded the older rocks in both the eastern and western parts of the island.

All of the known lodes in the region are on St. Lawrence Island; they include disseminated molybdenite in one of the plutons, a small low-grade porphyry copper deposit with minor molybdenite in a small satellitic stock, and several small sulfide deposits containing lead, zinc, and silver. None of these occurrences has been thoroughly explored. Anderson (1947, p. 41–42) mentioned a report of cassiterite near the southwestern end of the island but did not specify whether it was a bedrock or placer occurrence. Recent stream-sediment sampling and reconnaissance geologic mapping in the area failed to find any indication of tin mineralization (oral commun., Béla Csejtey, Jr., Sept., 1970). With the possible exception of the rumored cassiterite, no placer deposits have been reported from the land area of the Bering Sea region. There has been very little prospecting in the region, however, owing in part to its remoteness and in part to governmental restrictions.

Recent investigations in the Bering Sea (Nelson and Hopkins, 1969; Nelson and others, 1969) disclosed local concentrations of gold in bottom sediments, in particular between St. Lawrence Island and the Seward Peninsula. A little native copper of no probable economic interest was found in bottom samples collected near the northwest corner of St. Lawrence Island.

BRISTOL BAY REGION

The Bristol Bay region (pl. 1, fig. 4) includes the area drained by streams flowing into Bristol Bay from Cape Newenham on the

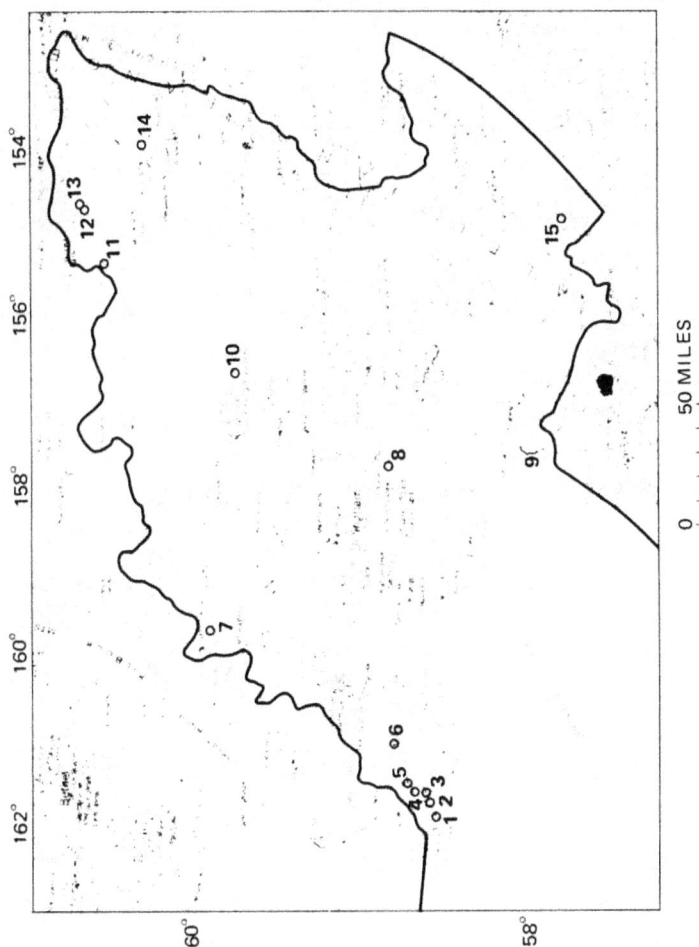

FIGURE 4.—Placer deposits in the Bristol Bay region.

1-6. Hagemeister Strait: Smith (1939a, p. 63),
      Berryhill (1963, p. 17-23).
7. Trail Creek: Hoare and Coonrad (1961).
8. Keefer's (Keeler's) Bar: Mertie (1938b, p.
      91; unpub. data.).
9. Eregik beach: Berryhill (1963, p. 28-30).
10. Mulchatna River (mouth of Stuyahok River):
      Unpub. data.
11. Lambert's Bar: Unpub. data.
12. Bonanza Creek: Jasper (1961). Scyrnneva
      Creek: Jasper (1961, p. 60-61, 64).
13. Bonanza Creek: Jasper (1961). Pass Creek:
      Jasper, (1961, p. 60-61).
14. Portage Creek: Capps (1935, p. 94).
15. Cape Kubugakli: Smith (1925, p. 206-207).

west to and including Egegik Bay on the east and into Shelikof Strait from Cape Douglas on the north to Cape Kekurnoi on the south. The region is considered as one district.

The southeastern part of the region consists of rugged mountains, the highest peaks of which are mainly Quaternary volcanoes, some active in historic time, with summits 5,000 to 7,500 feet in altitude. The northwestern part is a lake-dotted area less than 1,000 feet above sea level; isolated hills rise a few hundred to slightly more than 2,000 feet. The Ahklun Mountains in the western part of the region make up a low, but rugged range that contains large deep lakes of extraordinary scenic grandeur.

The mountains in the eastern part of the Bristol Bay region consist of rocks that range from possibly Permian metamorphosed volcanic rocks to Tertiary and Quaternary lava flows and fragmental rocks. The bulk of the bedded rocks are Mesozoic sandstone, shale, and conglomerate. Northwest of a major fault, the Bruin Bay fault, the older rocks were invaded and locally metamorphosed by the dioritic Aleutian Range batholith of Jurassic age and smaller younger felsic and mafic plutons and volcanic necks (Burk, 1965; Detterman and Reed, 1968).

In the western part of the region, bedrock is mainly Paleozoic and Mesozoic clastic and volcanic rocks and Tertiary felsic and mafic dikes, sills, and small plutons (Mertie, 1938b; Hoare and Coonrad, 1961). Between the eastern and western mountains, the region is a lowland underlain by thick glacial and alluvial deposits; bedrock is exposed only around its margins and in a few hills that protrude through the surficial materials. Except for its north-central part, the region was glaciated and is now mainly in zones characterized by isolated masses of permafrost.

Lode deposits containing mercury, gold, silver, copper, lead, zinc, antimony, and iron are known in the Bristol Bay region, but little ore has been produced from them (Berg and Cobb, 1967, p. 9–16, fig. 4).

No rich placer deposits have been found in the Bristol Bay region and few were ever developed much beyond the prospecting stage. As records of mining activity are almost nonexistent, even the locations of many of the reported occurrences of placer gold are open to question. Undoubtedly gold has been found in many more places than shown on the map (fig. 4). The total production of the region was probably at least 500 fine ounces, but not much more than 1,000.

Most of the gold probably came from Cape Kubugakli (15, fig. 4) and Portage Creek (14, fig. 4). At Cape Kubugakli a small, steep stream drains an area of numerous small sulfide-bearing quartz veins in fine-grained igneous rock. The best values in the creek were found immediately downstream from the veins. Portage Creek is about 5 miles long and enters Lake Clark from the northwest. From 1910 to 1912 and for a few years after World War II, some gold was recovered, but the total amount was probably worth only a few thousand dollars. Desultory mining and prospecting have been reported from other streams in the same general area, but there has been no activity on them for many years.

Bonanza Creek and its tributaries, Pass and Scynneva Creeks, (12, 13, fig. 4) have been extensively prospected, but production probably has been less than 150 fine ounces of gold. Quartz veins, some containing a few sulfide minerals and a little free gold, are the probable source of the gold in the creek gravels. The valley of Bonanza Creek, though narrow, might be capable of supporting a small dredge or a dragline operation under favorable economic conditions. The Nushagak River and some of its tributaries, particularly the Mulchatna River, are known to be auriferous and to have been the source of very small amounts of gold in the late 1800's and early 1900's. There was, however, no commercially successful mining in the Nushagak basin. Farther west, on Trail Creek (7, fig. 4), a headwater tributary of the Togiak River, there are signs of placer mining, but the results are not known.

A reconnaissance study of the U.S. Bureau of Mines (Berryhill, 1963) of beach sands around Bristol Bay failed to discover major concentrations of valuable minerals. Although an atypical sample from a beach south of Egegik (9, fig. 4) contained nearly 250 pounds of iron per cubic yard of beach material. There were traces of flour gold in a few samples from this beach and similar deposits on the northwest shore of Hagemeister Strait (1–6, fig. 4), where there was a small stampede in 1937 following overoptimistic reports by prospectors. Gold recovered from beach deposits around Bristol Bay was worth no more than a few hundred dollars. The beach gold probably was mainly reconcentrated from glacial deposits; some from Hagemeister Strait may have been derived from nearby sulfide-bearing veins.

COOK INLET-SUSITNA RIVER REGION

The Cook Inlet-Susitna River region (pl. 1, fig. 5) is the area drained by streams flowing into Cook Inlet between Cape Douglas on the south and Portage at the eastern end of Turnagain Arm. It comprises the Anchorage, Redoubt, Valdez Creek, Willow Creek, and Yentna districts.

The region includes much of the Alaska Range, which culminates at 20,300–foot Mount McKinley; the northern slopes of the western Chugach Mountains; and most of the Talkeetna Mountains. These areas are separated by poorly drained, lake-dotted lowlands, the most extensive of which extends northward from the head of Cook Inlet.

The oldest geologic units in the region are Paleozoic clastic and carbonate rocks, exposed mainly in the Alsaka Range. Mesozoic volcanic and clastic rocks, in which considerable limestone is interbedded locally, make up the bulk of the bedded rocks in the Chugach Mountains. Recent investigations in a neighboring area (Plafker and MacNeil, 1966) indicate that some of the clastic and interbedded volcanic rocks of the Chugach Mountains probably are early Tertiary in age. Large granitic batholiths of Jurassic, Cretaceous, and Tertiary age invaded the older sedimentary and volcanic rocks in the Talkeetna Mountains and Alaska Range (Dutro and Payne, 1957; Grantz and others, 1963; Reed and Elliott, 1970), and there are smaller plutons in the Chugach Mountains and other parts of the region. Part of a discontinuous belt of small ultramafic bodies of probable late Mesozoic age that extends from the southern tip of the Kenai Peninsula nearly to the Wrangell Mountains is in the Anchorage district.

Tertiary continental deposits underlie Cook Inlet and large areas in the Susitna Lowland and Matanuska Valley. Tertiary and Quaternary volcanic rocks have been found in the Matanuska Valley, west of Anchorage, and in the southern Alaska Range, where Augustine Island and some of the highest peaks are active volcanoes. Most of the lowlands are underlain by thick glacial and alluvial deposits. Except along the shore of Cook Inlet, in the lower Matanuska Valley, and in a belt on both sides of the Susitna River below Curry, most of the region is underlain by permafrost. Ice completely covered the Cook Inlet-Susitna River region during the Pleistocene, spreading from the Alaska Range far out to sea. Ice still covers the highest parts of the mountains, and valley glaciers extend many miles from their source areas; one of them, the Kahiltna Glacier, is more than 35 miles long.

Gold and silver have been recovered from lodes in many parts of the Cook Inlet-Susitna region and a little copper has been recovered from deposits in the Redoubt and Valdez Creek districts. These lodes and others have been investigated as possible sources of antimony, iron, chromite, molybdenum, copper, lead, and zinc (Berg and Cobb, 1967, p. 16–37, figs. 5–9; Reed and Eberlein, 1972). Reconnaissance studies in the southern Alaska Range in

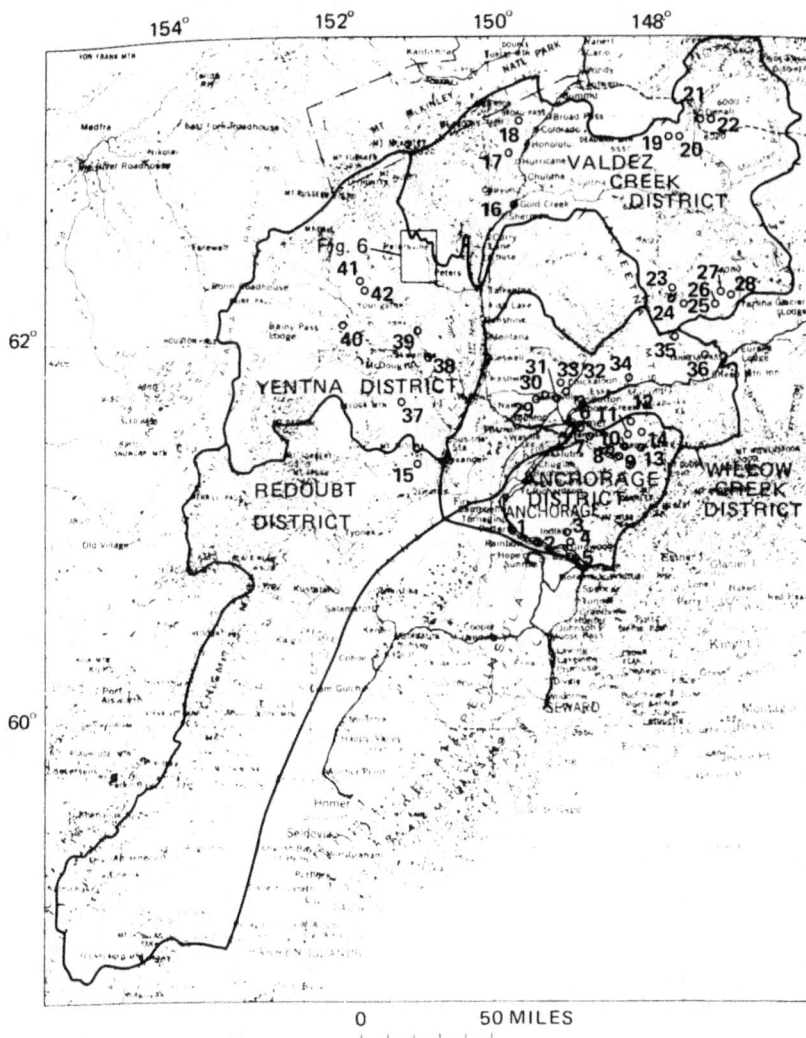

FIGURE 5.—Placer deposits in the Cook Inlet-Susitna River region.

Anchorage district

1. Potter Creek: Jasper (1967a, p. 31).
2. Indian: Capps (1916b, p. 187).
3. Crow Creek: Moffit (1960a, p. 41-43), Paige and Knopf (1907a, p. 121-122).
4. California Creek: Capps (1916b, p. 186). Jasper (1967a, p. 32). Crow Creek: Moffit (1906a, p. 40-43), Capps (1916b, p. 175-185). Winner Creek: Capps (1916b, p. 186).
5. Kern Creek: Capps (1916b, p. 186-187), Jasper, (1967a, p. 33). Peterson Creek: Jasper (1967a, p. 33).

6. McRoberts Creek: Jasper (1967a, p. 30). Unnamed occurrence: Jasper (1967a, p. 31).
7. Jim Lake: Jasper (1967a, p. 31).
8. Fall Creek: Richter (1967a, p. 16).
9-12. Metal Creek: Richter (1967a, p. 2. 8-10, 15-16).
13-14. Knik River, Glacier Fork: Richter (1967a, p. 15-16).

Redoubt district

15. Lewis River: Brooks (1918, p. 46-47), Alaska Department of Mines (1958, p. 63).

1969 (Reed and Elliott, 1970) indicate that metallic sulfide minerals are common in and near granitic plutons. Float samples collected in the vicinity of the Mount Estelle pluton in the southwestern part of the Yentna district contained as much as 60 parts per million (about 1.7 fine ounces per ton) gold associated with chalcopyrite, arsenopyrite, and other sulfides. The only large-scale production was from the Willow Creek area north of Palmer where, between 1909 and World War II, gold-bearing quartz veins in the southern border zone of the Talkeetna batholith were the source of about 404,425 fine ounces of gold, about 5 percent of Alaska's total lode gold output.

Placer gold was discovered in the Cook Inlet-Susitna region in the late 19th century, and mining has been continuous since the early 1900's. Total placer gold production from the region probably has been about 250,000 fine ounces, or less than 40 percent of the region's lode-gold production. Production from the Valdez Creek

---

FIGURE 5.—Continued.

Valdez Creek district

16. Susitna River: Capps (1919b, p. 231).
17. Shotgun Creek: Capps (1919b, p. 231).
18. Bryn Mawr Creek: Capps (1919b, p. 221, 231).
19. Gold Creek: Moffit (1912, p. 54).
20. Wickersham Creek: Moffit (1912, p. 54), Moffit (1915, p. 76).
21. Timberline Creek: Moffit (1912, p. 65). Valdez Creek: Ross (1933, p. 437, 444-453), Tuck (1938, p. 122-127), Smith (1970).
22. Lucky Gulch: Ross (1933, p. 454-455), Tuck (1938, p. 129). Rusty Creek: Moffit (1912, p. 64-65). White Creek: Ross (1933, p. 453-454), Tuck (1938, p. 128-129).
23. Granite Creek: Chapin (1918, p. 64).
24. Roaring Creek: Chapin (1918, p. 64).
25. Gold Creek: Chapin (1918, p. 64).
26. Yacko Creek: Chapin (1918, p. 64).
27. Fourth of July Creek: Chapin (1918, p. 64).
28. Daisy Creek: Chapin (1918, p. 64).

Willow Creek district

29. Grubstake Gulch: Capps (1915, p. 52-54). Willow Creek: Paige and Knopf (1907b, p. 66-67), Capps (1915, p. 52-55), Jasper (1962, p. 81-82).
30. Craigie Creek: Jasper (1967a, p. 23). Hatcher Creek: Jasper (1967a, p. 26). Upper Willow Creek: Jasper (1967a, p. 25-26).
31. Fishhook Creek: Capps (1915, p. 55).
32. Little Susitna River: Jasper (1967a. p. 27-29).

Willow Creek district—Continued

33. Reed Creek: Jasper (1967a, p. 26).
34. Chickaloon Creek (River): Mendenhall (1900, p. 322). Schoonoven (Boulder) Creek: Mendenhall (1900, p. 322).
35. Mazuma Creek: Martin and Mertie 1914, p. 278-279, 281), Brooks (1925, p. 30).
36. Alfred Creek: Martin and Mertie (1914, p. 278-279, 281), Brooks (1925, p. 30).

Yentna district

37. Texas Creek: Ralph Tuck (unpub. data, 1933).
38. Kahiltna River (Boulder Bench, Leslie's Bar, Red Hill Bar, Round Bend Bar): Mertie (1919, p. 263), Bates and Wedow (1953, p. 8), Robinson and others (1955, p. 22).
39. Kahiltna River (Sholan Bar): Mertie (1919, p. 262) Bates and Wedow (1953, p. 9).
40. Kichatna River: Martin (1919, p. 32-33).
41. Pass Creek: Smith (1939b, p. 39).
42. Big Boulder Creek: Capps (1913, p. 70). Chicago Gulch: Capps (1913, p. 68-69). John(s) Creek: Capps (1913, p. 70). Little Boulder Creek: Capps (1913, p. 70). Mills Creek: Capps (1913, p. 68-70), Alaska Department of Mines (1954, p. 39). Twin Creek: Alaska Department of Mines (1954, p. 39). Wagner Gulch: Capps (1913, p. 68).

district cannot be confidently separated from that of the neighboring Chistochina district of the Copper River region. The total for the two is somewhat less than 200,000 fine ounces, of which probably about 20 percent should be credited to the Valdez Creek district. The combined placer production of the Willow Creek and Yentna districts through 1960 was about 204,350 fine ounces; an unknown, but small amount has been produced since that time. As the main gold-producing part of the Anchorage district has historically been considered a part of the Kenai Peninsula region, its output cannot be stated accurately; probably it was not more than a few thousand ounces. Scant data suggest that not more than 275 ounces of gold was recovered in the Redoubt district, where a single stream was worked for a few years.

ANCHORAGE DISTRICT

The Anchorage district (pl. 1, fig. 5) is bounded on the south by Turnagain Arm, on the west and north by Knik Arm, and on the east by the divide between Cook Inlet and Prince William Sound.

Most of the placer mining in the district was on Crow Creek (4, fig. 5) near Girdwood. The creek, about 5 miles long, is fed by a glacier and drains an area where several gold-bearing lodes were explored and mined on a small scale. Bedrock is mainly interbedded slate and graywacke cut by numerous granitic dikes and sills. Crow Creek had a complex glacial and erosional history that involved the interaction of ice tongues that came down Crow Creek and Glacier Creek, the larger stream into which it flows. Any existing preglacial placers were destroyed and the material in them was scattered. Mining operations uncovered several old gravel-filled channels in the lower part of the creek. The gold was largely reconcentrated from glacial deposits; some may have been directly derived from lode sources uncovered after the ice fronts had retreated nearly as far as their present positions. Both coarse and fine gold and small amounts of native silver and copper are constituents of several pay streaks, the richest of which was at the base of the gravels and in the shattered top 2 feet of bedrock. Most mining was by hydraulic methods whereby all gravels excavated were run through a long string of sluice boxes. In the early 1900's an attempt was made to mine a gravel-filled basin behind the terminal moraine of a glacier that had come down a small tributary valley and temporarily dammed Crow Creek (3, fig. 5). California and Winner Creeks (4, fig. 5), also tributary to Glacier Creek, were mined from time to time with indifferent results.

Other streams that enter Turnagain Arm from the north generally resemble Crow Creek and were sites of prospecting and small-scale mining. Many reports of activity, however, are so vague that the places of actual operation could not be located closely enough to plot on the map (fig. 5).

Small amounts of gold, accompanied by a little platinum, were mined sporadically for many years on Metal Creek (9, fig. 5), but few data on the character of the deposits or on the production are available. Recent geochemical and geological reconnaissance in the Metal Creek area (Richter, 1967a) revealed scheelite in samples from Metal Creek (9–12, fig. 5), the Glacier Fork of Knik River (13–14, fig. 5), and Fall Creek (8, fig. 5). One sample contained native silver. Gold, restricted to the lower half of Metal Creek, probably was reconcentrated from glacial deposits, but the source of the platinum was not discovered. Richter concluded that, as gold has not been found in neighboring drainages, its ultimate source must be local and may be small quartz veins similar to those that occur in slate and graywacke near the mouth of Metal Creek.

A geochemical survey along highways in the Anchorage area by the Alaska Division of Mines and Minerals (Jasper, 1967a) revealed scheelite and various sulfide minerals in many concentrate samples. Cinnabar was identified in a sample from McRoberts Creek (6, fig. 5).

REDOUBT DISTRICT

The Redoubt district (pl. 1, fig. 5) is the area drained by streams flowing into Cook Inlet between Cape Douglas on the south and (but excluding) the Susitna River on the north.

The only successful placer mining in the district was on Lewis River (15, fig. 5), which heads in a swampy basin west of Mount Susitna. Below the basin the stream follows a valley in Jurassic or Cretaceous granitic rocks and a small patch of Tertiary conglomerate before it crosses a low coastal plain covered with Quaternary surficial deposits. Coarse gold worth perhaps $2,000 was recovered during prospecting in 1916–17, and an unknown but probably small amount was recovered in 1956–57. As some of the gold had fragments of quartz attached to it, the major source was probably small veins in the granitic rocks, although some gold may have been reconcentrated from Tertiary conglomerate.

In 1902 an unsuccessful attempt was made to recover gold from the Beluga River (Brooks, 1918, p. 45) southwest of Mount Susitna, but the location of this venture is not known. Bates and Wedow (1953, p. 8) reported monazite and zircon in placer samples from the Mount Spurr area in the northern part of the Redoubt district; the source and mode of occurrence of the material represented by the samples are unknown.

VALDEZ CREEK DISTRICT

The Valdez Creek district (pl. 1, fig. 5) is the area drained by the Susitna River above the mouth of the Talkeetna River and includes the Chulitna River basin.

The most extensive placer mining in the districts, and the only place where there has been significant activity since World War II, is at Valdez Creek (21, fig. 5), which drains an area underlain by metamorphosed sedimentary and volcanic rocks and by small felsic and mafic intrusive masses. Several small gold- and sulfide-bearing quartz veins are in the area, but only a small amount of ore was mined from them. These lodes, and similar ones that may now be buried beneath surficial deposits, were the probable source of the gold in the placers of Valdez Creek and its tributaries.

Valdez Creek has a complex Pleistocene history marked by changes in local base level brought about by the interaction of small local glaciers and a large ice tongue in the Susitna Valley. Deposits in old channels discovered where the present course of the creek crosses them were the source of most of the gold mined, although the modern stream gravels also are auriferous. Recent work (T. E. Smith, 1970) demonstrated that not all of the old channels have been mined out and that a large volume of bench gravels near the mouth of Valdez Creek probably has a gold value of 50 cents to $1.20 per cubic yard. Using minimum tenor and volume figures, Smith estimated the potential value of the gold resources of the bench gravels (including unmined parts of old channels) to be more than $17 million (gold at $35 per ounce). Most of the mining on Valdez Creek was by drifting in buried channels and by hydraulic methods. Smaller scale operations were carried on intermittently in stream and bench gravels on White Creek (22, fig. 5), Lucky Gulch (22, fig. 5) and other tributaries of Valdez Creek. Minerals identified in concentrate samples in addition to gold include magnetite, pyrite, zircon, sphene, sillimanite, kyanite, galena, realgar, orpiment, and hessite, a silver telluride. Small amounts of gold were found in Gold (19, fig. 5), Wickersham (20, fig. 5) and other creeks.

Although there are lode deposits in the area drained by the West Fork of the Chulitna River in the north-central part of the Valdez Creek district, including one that produced gold and copper ore, no valuable placers have been found. In the early 1900's a little gold was recovered from Bryn Mawr (18, fig. 5) and Shotgun (17, fig. 5) Creeks, but the results evidently were discouraging, as not even prospecting has been reported for many years.

Most of the bars of the Susitna River below Gold Creek (16, fig. 5) carry minor concentrations of flour gold at or near the surface.

In the southeastern part of the Valdez Creek district, a little gold is in the gravels of several creeks. Daisy Creek (28, fig. 5), which crosses a contact between Jurassic volcanic and coarse clastic rocks, was worked on a small scale at several places before 1914.

Placer deposits in the western part of the Valdez Creek district are geologically akin to neighboring deposits in the Yentna district and are discussed with them.

WILLOW CREEK DISTRICT

The Willow Creek district (pl. 1, fig. 5) is the area drained by eastern tributaries of the Susitna River below Sunshine, by northern tributaries of Cook Inlet and Knik Arm east of the Susitna, and by the Matanuska River.

Most of the placer gold from the district was recovered from streams that drain areas containing once-productive lode mines. Considerably more than half of the placer gold probably came from operations on Grubstake Gulch and the part of Willow Creek immediately below the mouth of Grubstake Gulch (29, fig. 5), where claims were staked as early as 1897 and mining was still in progress in 1969. The gold in the gulch was derived from auriferous quartz veins in mica schist, one of which was worked successfully at the Thorpe lode mine, near the head of a fork of Grubstake Gulch (Berg and Cobb, 1967, p. 31, 34). In other streams in the same general area, placer gold derived from veins in a dioritic batholith was found but not mined. Jasper (1967a), in the course of a geochemical reconnaissance, collected many concentrates from this part of the Willow Creek district and identified scheelite and chalcopyrite in several samples.

On Alfred Creek (36, fig. 5), workable gold placers were discovered in 1911 and mined sporadically on a small scale for many years. A little platinum accompanies the gold. Minor production of gold was reported from upper Mazuma Creek (35, fig. 5), a nearby stream that drains an area underlain only by volcanic rocks and Tertiary conglomerate. According to Martin and Mertie (1914, p. 279–280) and Grantz (1956), the gold in the area must have been reconcentrated from the conglomerate, which originally covered a much greater area than the remnants preserved. Mendenhall (1900, p. 321–322) noted that gold had been found on several south-flowing tributaries of the Matanuska River, but later reports did not describe successful mining on any of them.

YENTNA DISTRICT

The Yentna district (pl. 1, fig. 5) includes the area drained by western tributaries of the Susitna River between Alexander and Sunshine and by its eastern tributaries from Sunshine to and including the Talkeetna River.

Most of the placer mining in the district was in the Cache Creek area on streams draining the Peters and Dutch Hills (fig. 6) and

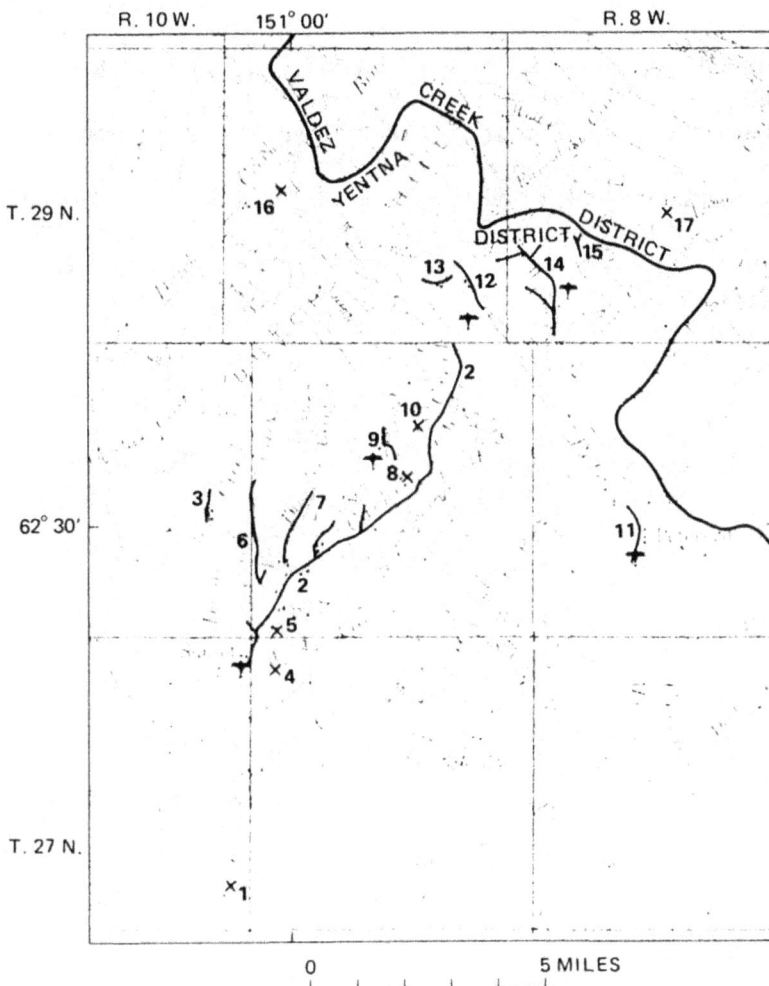

FIGURE 6.—Placer deposits in the Cache Creek area.

near Fairview Mountain (41, 42, fig. 5) about 20 miles to the southwest.

The Peters and Dutch Hills are largely underlain by graywacke and finer clastic rocks, predominantly Mesozoic in age. Tertiary continental rocks, including auriferous conglomerate, occur in the Dutch Hills and are exposed in many creek valleys. Alaskite dikes and small plutons and at least one mafic or ultramafic dike, now altered to silica-carbonate rock, cut the Mesozoic sequence (Clark and Hawley, 1968). Some of the dikes and many quartz veins in the Mesozoic rocks contain a little magnetite, scheelite, various sulfide minerals, and native gold, but none of these occurrences has been developed into a mine. Except in parts of the Dutch and Peters Hills, Quaternary glacial and alluvial deposits blanket bedrock in most of the area. Placers include stream and bench deposits of the present streams, glaciofluvial deposits of Pleistocene age, and Tertiary conglomerates. On the basis of recent work by Clark and Hawley (1968), auriferous quartz-rich conglomerates and breccias on Dollar (3, fig. 6), Thunder (7, fig. 6), and Willow (14, fig. 6) Creeks, once considered to be extensively weathered deposits that were buried on an old erosion surface, are now thought to be the result of erosion of hydrothermally altered zones that follow northeastward-striking steep faults.

Most of the gold mined in the Cache Creek area probably came from dredging operations on Cache (2, fig. 6) and Peters (11, fig. 6) Creeks. Streams draining the southeast flank of the Dutch Hills were extensively mined by hand and nonfloat methods. Falls

---

FIGURE 6.—Continued.

1. Cache Creek: Smith (1938, p. 42–43).
2. Cache Creek: Capps (1913, p. 53–57), Mertie (1919, p. 243–248), Robinson and others (1955, p. 2). Lucky Gulch: Capps (1913, p. 60). Rambler Creek: Capps (1913, p. 57). Short Creek: Smith (1933a, p. 28).
3. Dollar Creek: Mertie (1919, p. 252–254), Capps (1925, p. 54–57, 59), Clark and Hawley (1968, p. 35–40).
4. Windy Creek: Mertie (1919, p. 254).
5. Cheechako Gulch: Smith (1933b, p. 29).
6. Falls Creek: Mertie (1919, p. 251–252). Ruby Creek (Gulch): Smith (1939b, p. 39). Treasure Creek: Smith, S. S. (1917, p. 42).
7. Thunder Creek: Mertie (1919, p. 249–251), Capps (1925, p. 55, 57–59), Clark and Hawley (1968, p. 35–40).
8. Nugget Creek: Capps (1913, p. 60).
9. Nugget Creek: Capps (1913, p. 58–60), Mertie (1919, p. 248–249).
10. Gold Creek: Capps (1913, p. 58).
11-12. Peters Creek: Capps (1913, p. 64–65), Mertie (1919, p. 255–257).
13. Bird Creek: Mertie (1919, p. 260–261).
14. Falls Gulch: Capps (1913, p. 66). Gopher Creek (Gulch): Mertie (1919, p. 260). Rocky Gulch: Capps (1913, p. 66). Ruby Creek (Gulch): Mertie (1919, p. 259–260). Slate Gulch: Capps (1913, p. 66). Snow Gulch: Capps (1913, p. 66). Willow Creek: Mertie (1919, p. 259–260), Capps (1925, p. 55, 58–59), Clark and Hawley (1968, p. 36–40).
15. Poorman Creek: Mertie (1919, p. 257–259), Robinson and others (1955, p. 2).
16. First Creek: Smith (1938, p. 43).
17. Canyon Creek: Mertie (1919, p. 261–262).

(6, fig. 6), Thunder, Nugget (8, 9, fig. 6), Bird (13, fig. 6), and Willow Creeks and their tributaries have been major producing streams; some have been worked as recently as 1967. Gold has been found in an adjacent part of the Valdez Creek district where several streams tributary to the Tokositna (Tokichitna) River (fig. 5) drain the northeastern end of the Dutch Hills. Mining was reported from Canyon Creek (17, fig. 6) and on Ramsdyke and other nearby creeks at places that cannot be accurately located. On the northwestern flank of the Dutch Hills there has been profitable mining on First Creek (16, fig. 6), but none on any of the nearby streams. The only creek rising in the Peters Hills that was a site of mining is Windy Creek (4, fig. 6), where a bench deposit of glaciofluvial material was worked.

Concentrates from placers in the Cache Creek area contain a greater variety of heavy minerals than those from most other parts of Alaska. Platinum, possibly derived from bedrock sources such as an altered mafic or ultramafic dike described by Clark and Hawley (1968, p. 17–18), has been identified in samples from Cache Creek and many of its tributaries, from Poorman (15, fig. 6) and Willow Creeks in the upper basin of Peters Creek, and from Canyon Creek. Cassiterite and scheelite are widespread and native copper and various sulfide minerals have been found in many creeks. Uranium and thorium were identified by analyses of samples from Cache and Poorman Creeks.

The area around Fairview Mountain has not been as extensively studied as that near Cache Creek. Poorly consolidated Tertiary sandstone and conglomerate underlie the part of the area where there has been mining, and older metasedimentary rocks are exposed on the north flank of Fairview Mountain (Barnes, 1966, pl. 2). Mills and Twin Creeks and their tributaries (42, fig. 5), which rise on the southeast flank of the mountain and join soon after reaching the lowland, were mined nearly all the way from their confluence upstream into headwater gulches. Gold was discovered in Wagner Gulch (42, fig. 5), a tributary of Mills Creek, in 1905, and mining was carried on in the area until the middle 1950's. As the operations were eventually integrated on a fairly large scale, the amount of gold produced must have been large, possibly rivalling that of the Cache Creek area. Pass Creek (41, fig. 5) also was the source of significant production. Platinum was reported from the basin of Lake Creek (Martin, 1919, p. 33), the stream into which the creeks around Fairview Mountain drain, but this occurrence cannot be located precisely.

The Kahiltna River receives all drainage from the Cache Creek and Fairview Mountain areas. Gold production from bars of this stream (38, 39, fig. 5) was reported for many years, but the total amount was probably small. Of particular interest is the fact that fine platinum was recovered as well as fine gold. Other heavy minerals in the concentrates included cassiterite, scheelite, magnetite, monazite, uranothorianite, rutile, and garnet. The heavy minerals may have been reconcentrated from Tertiary rocks exposed at Sholan Bar (39, fig. 5) and near other bars (38, fig. 5) where there was mining, or they may have derived from Quaternary deposits. They may even have come from as far away as Cache Creek.

Prospecting in other tributaries of the Yentna River failed to find workable placers, although gold was discovered in several places. In 1933, Ralph Tuck (U.S. Geol. Survey, unpub. field notes) visited Texas Creek (37, fig. 5), were 18 claims had been staked. He reported that a shallow trench 200 feet long had reached a false bedrock of the blue clay and that a few fine colors of gold had been found but that the grade of the deposit was low. In 1917, placer-mining machinery was installed on the Kichatna River at the mouth of the Nakochna River (40, fig. 5), but there is no record that it was operated successfully. In addition to fine gold, platinum was reported from the Kichatna River. Other reports of placer gold in the Yentna River basin are so vague that the places where there was prospecting cannot be located within many miles.

COPPER RIVER REGION

The Copper River region (pl. 1, fig. 7) includes the area drained by the Copper River and its tributaries, the area east of the divide between Prince William Sound and Cook Inlet, the area drained by streams flowing into the Gulf of Alaska between the Copper River and long 141° W., and offshore islands, including Middleton Island. It is divided into five districts: Chistochina, Nelchina, Nizina, Prince William Sound, and Yakataga.

The region includes parts of the Alaska Range, Wrangell, St. Elias, and Kenai-Chugach Mountains and extensive lowlands along the Copper and Chitina Rivers. The mountains, which rise to summits more than 16,000 feet in the Wrangells, and to more than 18,000 feet at Mount St. Elias, support and nourish the largest icefields and piedmont ice lobes and some of the longest valley glaciers in North America, all remnants of even more extensive Pleistocene ice that covered most of the region. The lowlands along the Copper River are floored by thick accumulations of Pleistocene and Holocene glacial, lacustrine, and fluvial deposits that are frozen to depths of several hundred feet. The islands and most of the shores of Prince William Sound, the extensive Copper

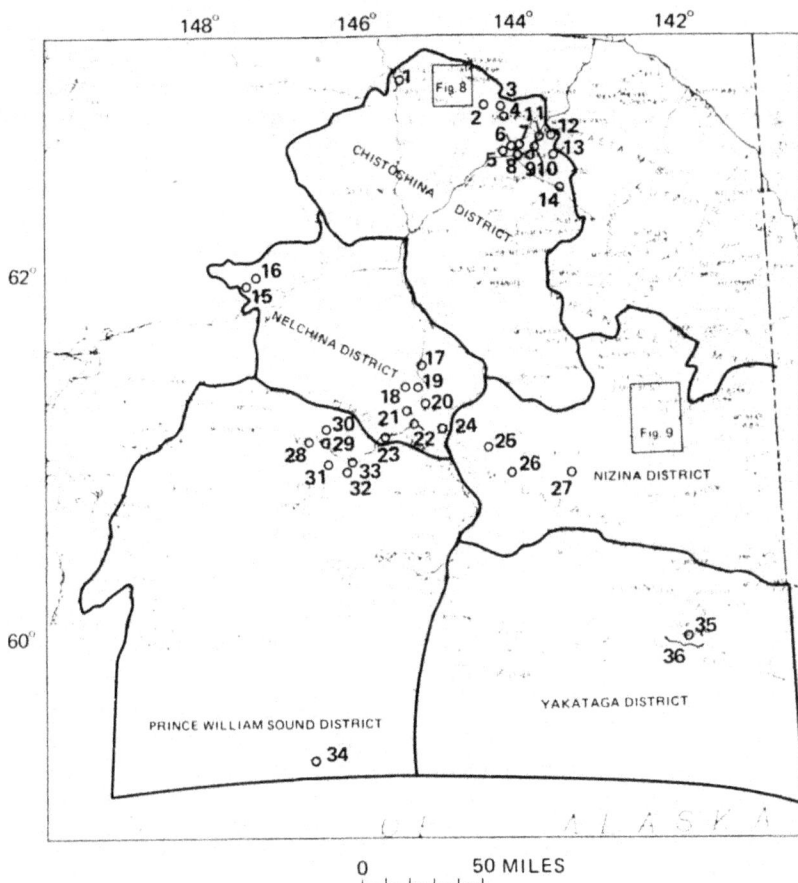

FIGURE 7.—Placer deposits in the Copper River region.

Chistochina district

1. Gunn Creek: Rose and Saunders (1965, p. 15).
2. Eagle Creek: Moffit (1944, p. 40–42).
3. Slana River tributary: Richter (1967b, p. 16).
4. Granite Creek: Moffit (1938b, p. 51), Moffit (1954a, p. 196).
5. Hidden Creek: Moffit (1938b, p. 51).
6. Grubstake Creek: Moffit (1938b, p. 48–50), Moffit (1954a, p. 195), Richter and Matson (1968, p. 2–3).
7. Boulder Creek: Richter (1966, p. 34). Cottonwood Creek: Matson (1969b, p. 3). Slope Creek (tributary to Porcupine Creek): Moffit (1938b, p. 50–51), Moffit, 1944, p. 43–44), Richter and Matson, 1968, p. 2–3).
8. Ahtell Creek: Moffit (1938b, p. 51). Willow Creek: Richter (1966, p. 34).
9. Porcupine Creek: Matson (1969b, p. 3).
10. Carlson Creek: Matson (1969b, p. 3).

Chistochina district—Continued

11. Unnamed creek: Matson (1969b, p. 3).
12. Bear Valley: Matson (1969b, p. 3).
13. Upper Suslota Lake: Matson (1969b, p. 3).
14. Unnamed creek: Matson (1969b, p. 3).

Nelchina district

15. Albert Creek: Chapin (1918, p. 59–62), Martin (1920, p. 23). North Creek: Martin and Mertie (1914, p. 278). Poorman Creek: Chapin (1918, p. 62). South Creek: Martin and Mertie (1914, p. 278).
16. Crooked Creek: Chapin (1918, p. 60–61).
17. Little Tonsina River: Jasper (1967b, p. 17).
18. Quartz Creek: Mendenhall (1905, p. 121), Moffit (1918, p. 179).
19. Fourth of July Creek: Smith (1932, p. 28).
20. Fall Creek: Moffit (1918, p. 181–182), Moffit (1938a, p. 127). Unnamed locality: Jasper (1967b, p. 15).
21. Boulder Creek: Moxham and Nelson (1952, p. 3).

River Delta, and the lowlands and low mountains bordering the Gulf of Alaska in the Yakataga district are generally free of permafrost.

The following summary of the geology of the Copper River region is based mainly on reports and maps by Brabb and Miller (1962), Jones and MacKevett (1969), MacKevett and Smith (1968), Moffit (1938a, 1954a, 1954b), Plafker (1967), Plafker and MacNeil (1966), and Smith and MacKevett (1970).

Bedrock in the Copper River region ranges in age from late Paleozoic to Quaternary. The bulk of the rocks are of Mesozoic age and include large masses of graywacke, slate, and greenstone and lesser amounts of carbonate rocks. Recent work has shown that some of the rocks near Prince William Sound previously considered to be Mesozoic are Tertiary in age. In late Mesozoic and early Tertiary time, plutons, some of batholithic dimensions, were emplaced in many parts of the region. They range in composition from granite to dunite but most are granodiorite, quartz diorite, and related rock types.

In the Yakataga district, complexly deformed Cenozoic marine and continental rocks underlie the area between the crest of the Chugach Mountains and the Gulf of Alaska and may be continuous with similar coeval rocks in Cook Inlet and on Kodiak Island. Middleton Island is composed of slightly indurated marine clastic sediments that were deposited in part by floating ice and are correlative with generally similar rocks exposed on the mainland. The most extensive formation in the Wrangell Mountains is a thick pile of Tertiary and Quaternary basaltic flows and associated rocks. The crater of Mount Wrangell (14,005 ft) still emits steam and ash.

Lodes in many parts of the Copper River region contain copper, gold, silver, molybdenum, antimony, nickel, chromite, lead, and zinc, but only copper, gold, and byproduct silver were mined

FIGURE 7.—Continued

Nelchina district—Continued

22. Stuart Creek: Jasper (1967b, p. 15).
23. Ptarmigan: Jasper (1967b, p. 10). Worthington Glacier: Jasper (1967b, p. 10).
24. Tiekel River: Brooks (1914, p. 62).

Nizina district

25. Little Bremner River: Moffit (1914, p. 48–49).
26. Bremner River (Threemile Canyon): Moffit (1914, p. 47–48).
27. Golconda Creek: Moffit (1914, p. 43–47).

Prince William Sound district

28. Gold Creek: Johnson (1915, p. 159).

Prince William Sound district—Continued

29. Mineral Creek: Grant and Higgins (1910, p. 72).
30. Mineral Creek: Johnson (1915, p. 186).
31. Solomon Gulch: Schrader (1900, p. 421).
32. Sulphide Gulch: Rose (1965b, p. 13-14).
33. Lowe River: Jasper (1967b, p. 7).
34. Middleton Island: Brooks (1913, p. 43).

Yakataga district

35. White River: Maddren (1914, p. 138-141).
36. Yakataga Beach: Maddren (1914, p. 134-138, Thomas and Berryhill (1962, p. 7, 16-17, 19-20).

commercially (Berg and Cobb, 1967, p. 37–73, figs. 10–13). The famous Kennecott mines near McCarthy in the Nizina district and mines in the southwestern and northeastern parts of Prince William Sound accounted for most of the copper produced in Alaska. Gold worth $2 or $3 million and smaller amounts of silver were produced from mineralized quartz and calcite veins and byproducts of copper mining in the Prince William Sound district. Similar veins near Golconda Creek (27, fig. 7) and in the southeastern part of the Nelchina district were mined on a small scale, but the entire region was not a statistically significant contributor to the total lode-gold production of Alaska.

Placer deposits have been worked in all districts of the Copper River region, but the total production probably was no more than 350,000 fine ounces of gold and a few ounces of platinum. Placers near the head of the Chistochina River and near Slana in the northern and northeastern parts of the Chistochina district accounted for an estimated 150,000–160,000 ounces of gold and all the platinum; deposits in the north-central part of the Nizina district accounted for about the same amount of gold; and beach and stream placers in the Yakataga and Nelchina districts, practically all the remainder. Placer-gold production from the Prince William Sound district probably did not exceed 500 ounces.

CHISTOCHINA DISTRICT

The Chistochina district (fig. 7) is the area drained by the Copper River and its tributaries above Gulkana and that part of the Copper River basin bounded on the west by the Copper River between Gulkana and Chitina and on the south by the Chitina River between Chitina and the mouth of the Nizina River.

Most of the placer mining in the district was in the extreme headwaters of the Chistochina River and its Middle Fork (fig. 8), where bedrock consists mainly of upper Paleozoic bedded rocks and probably Mesozoic mafic, ultramafic, and dioritic plutons. Older metamorphic rocks are exposed north of the Denali fault, a major regional tectonic feature here covered by the Chistochina Glacier. Gold-bearing Tertiary continental deposits, mainly poorly consolidated conglomerate and sandstone, are preserved between the head of Miller Gulch (4, fig. 8) and Big Four Creek (3, fig. 8) and in small patches in the valley of Slate Creek (4, fig. 8).

Gold was discovered on the Chisna River (11, fig. 8) in 1898. Mining in the area began the next year and continued into the 1960's. Most of the production was from rich stream gravels on Miller Gulch and Slate Creek. Gold was also mined for many years from Big Four Creek when water was available, from small

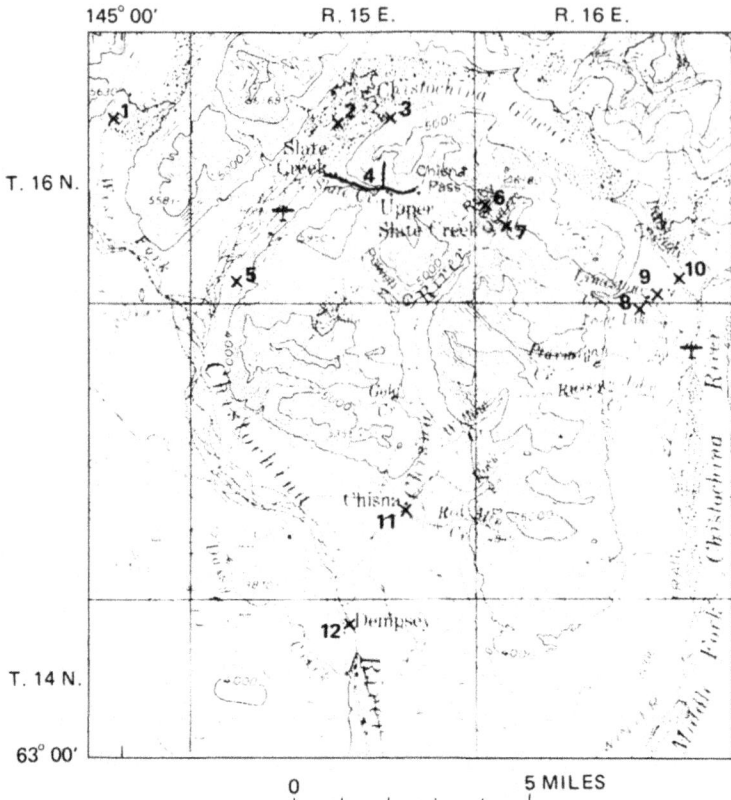

FIGURE 8.—Placer deposits in the upper Chistochina River area.

1. Chistochina River, West Fork: Rose (1967, p. 35–36).
2. Chistochina River: Rose (1967, p. 26).
3. Big Four Creek: Rose (1967, p. 25–26).
4. Miller Gulch: Moffit (1954a, p. 191–193), Rose (1967, p. 23–25). Slate Creek: Moffit (1944, p. 31–33), Rose (1967, p. 23–25).
5. Chistochina River: Rose (1967, p. 35–36).
6. Ruby Gulch: Moffit (1954a, p. 191–192).
7. Quartz Creek: Rose (1967, p. 26).
8-10. Chistochina River, Middle Fork (Bedrock Creek, Kraemer Creek, Limestone Creek): Moffit (1944, p. 34–40).
11. Chisna River: Moffit (1944, p. 29–31).
12. Dempsey: Moffit (1912, p. 77).

creeks at the head of the Middle Fork of the Chistochina River (8–10, fig. 8), and from several other streams. The gold in the richest placers was reconcentrated from both Tertiary conglomerate (called "round wash" by local miners) near the head of Miller Gulch and glacial deposits. The gold in most other placers, including a prominent bench on the north side of Slate Creek, probably was derived from glacial deposits. As no local bedrock source of gold has been found, several theories as to the ultimate source have been advanced. Rose (1967, p. 24–25) summarized them and concluded that the most probable source was undiscovered lodes north of the Denali fault, possibly near Mount Kimball, which may have been closer to the headwaters of the Chistochina River before the rocks were faulted.

Platinum, in the ratio to gold of 1:100, was recovered from concentrates from Slate Creek, Miller Gulch, and the head of the Middle Fork of the Chistochina River. It probably came from some of the ultramafic bodies in the area. Other heavy minerals identified in concentrate samples include magnetite, pyrite, chromite, native copper and silver, galena, cinnabar, and garnet. Scheelite was reported in concentrates from the Chistochina River (5, fig. 8) and its West Fork (1, fig. 8).

A little gold was mined from Eagle Creek (2, fig. 7) a few miles east of the Slate Creek area in 1942. Native copper, platinum, magnetite, barite, and other heavy minerals occur with the gold. Gold has been found on an unnamed tributary of the Slana River (3, fig. 7) and on Granite Creek (4, fig. 7), apparently in unprofitable amounts. A concentrate sample from Gunn Creek (1, fig. 7) near Isabell Pass 20 miles west of Slate Creek contained magnetite, chromite, ilmenite, zircon, scheelite, gold, and sphalerite, but not in minable amounts.

From 1934 until the late 1950's, there was small-scale placer-gold mining on Grubstake (6, fig. 7) and Slope (7, fig. 7) Creeks near the village of Slana. The area is underlain mainly by late Paleozoic bedded rocks that were intruded by a large diorite-quartz diorite complex and by a zoned quartz monzonite and granodiorite pluton, both of early Mesozoic age. Lode deposits associated with the zoned pluton contain galena and other sulfide minerals; several have been explored but have not been developed into mines. Gold is probably related to the diorite-quartz diorite complex (Richter and Matson, 1968). Total production from Grubstake and Slope Creeks probably was not more than 1,000 ounces. In addition to gold, concentrates contained magnetite, ilmenite, pyrite, bismuth, and native copper and silver, a mineral assemblage strikingly similar to that reported from streams in the upper Chistochina River area. Gold is known to be present in the gravels of many other streams in the Slana area (5, 7–14, fig. 7), but on none was activity carried beyond the stage of prospecting.

NELCHINA DISTRICT

The Nelchina district (fig. 7) is the area drained by east-flowing tributaries of the Copper River from Gulkana on the north to (but excluding) the Tasnuna River on the south.

In the northwestern part of the district, gold was discovered in 1912 on Albert Creek (15, fig. 7), setting off a small stampede that died out in a year or two. Valuable deposits were not found on any other creek in the area, but small-scale mining continued intermittently on Albert Creek until 1961 at least. The area is underlain by Mesozoic volcanic and marine sedimentary rocks in which lode occurrences of gold have not been found. Fine gold occurs nearly everywhere in the glacial and glaciofluvial deposits that cover much of the area but not in minable concentrations. Chapin (1918, p. 60) considered these deposits to be the major source of the gold in Albert and other creeks. Grantz (1956), however, considered a more probable source to have been an extensive cover of Tertiary terrestrial deposits, only small remnants of which are preserved. Geologic relations at nearby Mazuma Creek (in the Willow Creek district) and the size, shape, and composition of gold from Albert Creek are not consistent with an origin in local bedrock or glacial drift. The deposit mined at Albert Creek consisted of about 5 feet of poorly stratified coarse gravel containing many graywacke boulders. Gold was fairly evenly distributed through the gravel with no apparent concentration on bedrock. The material handled in 1913 averaged more than $10 a cubic yard. A little platinum accompanied the gold.

In the southeastern part of the district, a small amount of gold was mined from Fall Creek (20, fig. 7) and other streams. The streams head in Mesozoic slate and graywacke containing small auriferous quartz veins (Berg and Cobb, 1967, p. 48–49) that probably were the source of most of the placer gold. Some gold may have been reconcentrated from glacial deposits. Fall Creek flows, for most of its extent, in a narrow glaciated valley. Gold has been found in both stream gravels and low benches; most of it is concentrated in cracks in the upper foot or two of bedrock and on the bedrock surface. The presence of placer gold in this general area was known before 1900, for Schrader (1900, p. 422), who passed through in 1898, reported that gold had been found on several streams. In 1898–99 a placer deposit on Quartz Creek (18, fig. 7) was mined out. Development work on Fourth of July Creek (19, fig. 7) in 1929–30 is the most recent placer mining activity reported. During a geochemical reconnaissance along the highway from Valdez to Chitina in 1966, Jasper (1967b) identified scheelite in concentrates of samples collected between Worthington Glacier (23, fig. 7) and the Little Tonsina River (17, fig. 7) where it enters the Copper River valley.

NIZINA DISTRICT

The Nizina district (fig. 7) is the area drained by eastern tributaries of the Copper River between Chitina and Miles Glacier, excluding the area drained by northern tributaries of the Chitina below the Nizina River.

Most of the successful placer mining in the district has been on Dan (1, fig. 9) and Chititu (3, fig. 9) Creeks and their tributaries. The area is underlain by Triassic volcanic rocks and limestone and a thick section of Cretaceous clastic rocks, and by small Tertiary felsic and intermediate plutons (MacKevett and Smith, 1968, p. 1–3). Quartz veins in the plutons and in contact zones in the adjoining Cretaceous rocks carry various sulfide minerals and gold. Native copper occurs in the Triassic volcanic rocks, where some is localized between flows and some forms amygdules. The copper was probably an original constituent of the volcanic rocks, but some may have been introduced by hydrothermal solutions along fault zones. The placer deposits were formed by the erosion of the veins and volcanic rocks and by reconcentration of heavy minerals from Quaternary valley fills, remnants of which are preserved as benches high above the present streams. Heavy minerals in placer-concentrate samples collected in the area include gold, native copper and silver and mixtures of the two in individual nuggets, galena, stibnite, and lead. Much of the lead was probably artificially introduced, but some may be an original constituent of the deposits.

Gold was found on Dan Creek in 1901 and on Chititu Creek the next year, setting off a stampede that resulted in much claim staking and the establishment of a shortlived town on Chititu Creek. Young Creek (5–7, fig. 9) was staked at about the same time, but large-scale mining did not develop there or on neighboring Canyon Creek (8, fig. 9), where gold was found on river bars near its mouth and on rim rock in a canyon. Dan Creek has one of the longest essentially continuous histories of mining in Alaska; it was still being worked in 1968. In addition to gravel of the present stream, old channels and bench deposits have been mined. Native copper, about 40 tons of which was saved and sold, makes up a large part of the concentrates. Of particular interest was the discovery, in 1939, of a single copper nugget that weighed an estimated 3 tons. Chititu Creek and its major tributary, Rex Creek (3, fig. 9), are similar to Dan Creek, both in the types of placer deposits and in history of mining, although there is no record that any of the abundant copper was ever marketed.

The only other stream in the Nizina district that was the site of successful placer mining is Golconda Creek (27, fig. 7), where

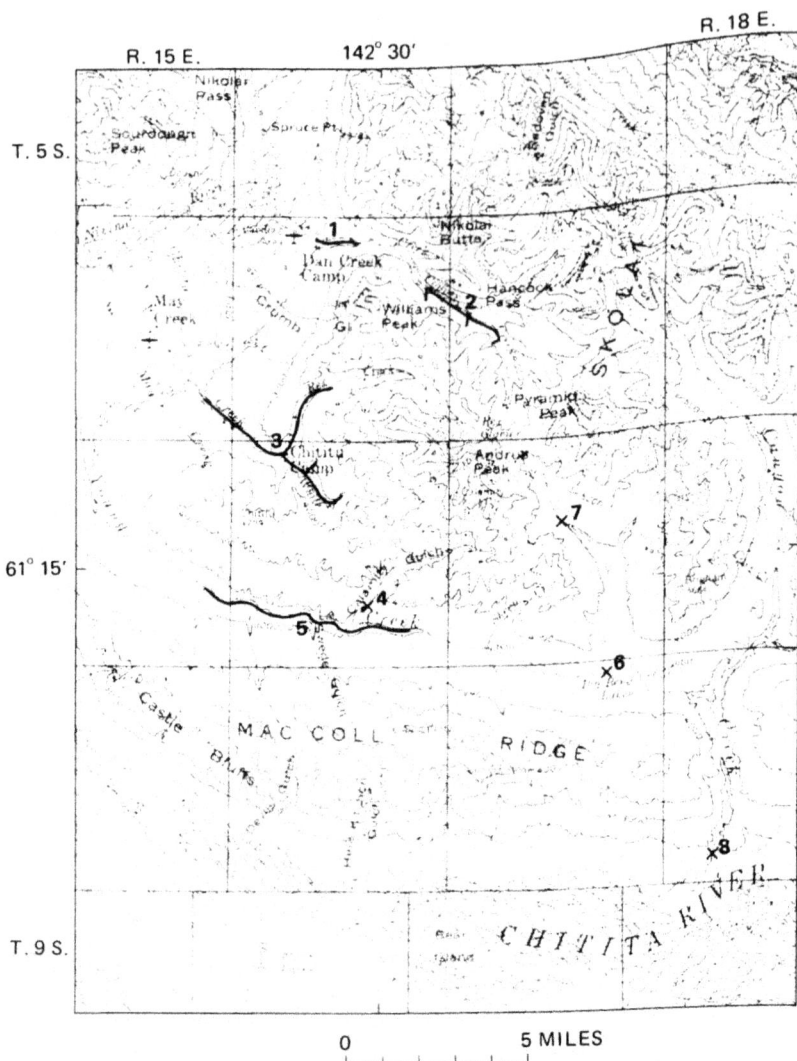

FIGURE 9.—Placer deposits in the Dan Creek-Young Creek area.

1. Dan Creek: Moffit and Capps (1911, p. 80, 98–113), Smith (1939a, p. 38–39), Miller (1946, p. 119–120).
2. Copper Creek and Idaho, Rader, and Seattle Gulches: Moffit and Capps (1911, p. 100–101).
3. Chititu Creek: Moffit and Capps (1911, p. 98–100, 103–107). Jolly Gulch: Brooks (1914, p. 62). Rex Creek (Gulch): Moffit and Capps (1911, p. 98–100, 103–107), Brooks (1915, p. 17). White Creek: Moffit and Capps (1911, p. 103–107).
4. Calamity Gulch: Moffit and Capps (1911, p. 108).
5-7. Young Creek: Moffit and Capps (1911, p. 107–108), Miller (1946. p. 98).
8. Canyon Creek: Moffit (1916, p. 135).

gold was discovered in 1901 and mined during about the next 15 years. The gold was derived from auriferous quartz veins, some of which were mined in the 1930's and 1940's, in probably Cretaceous slate cut by numerous light-colored fine-grained dioritic dikes.

A little fine gold can be panned from practically all of the other streams draining the north flank of the Chugach Mountains, but no workable placers have been discovered. In the early 1900's there were attempts to mine glaciofluvial terrace deposits and stream gravel derived from them on the Bremner (26, fig. 7) and Little Bremner (25, fig. 7) Rivers, but the amount of gold recovered was small and the enterprises were abandoned within a few years. Moffit (1916, p. 135) reported attempts at placer mining in the headwaters of the Kiagna River, the next major tributary of the Chitina above the Tana River. The Kiagna River drains a partially unexplored glaciated area where granitic rocks intrude Mesozoic and possibly Tertiary altered volcanic and clastic rocks.

PRINCE WILLIAM SOUND DISTRICT

The Prince William Sound district (fig. 7) comprises the area drained by streams flowing into Prince William Sound and the Gulf of Alaska from Cape Junken on the west to the Glacier River on the east and the area drained by western tributaries of the Copper River below and including the Tasnuna River. The Copper River Delta and Middleton Island are in the district.

Although major copper and gold lode mines were operated in the Prince William Sound district, no valuable placer deposits have been found in this intensely glaciated area. Small operations were carried on from time to time, mainly before World War I, on small streams that drain areas with once-productive lode mines near Valdez. Gold has been found in streams near Port Nellie Juan and in float at Jackpot Bay near Chenega in the western part of the district (Grant, 1909). Attempts to recover gold from the Lowe River east of Port Valdez (Moffit, 1954b, p. 308) were unsuccessful. During a geochemical reconnaissance in 1966, Jasper (1967b) identified scheelite in concentrate samples from near the mouth of the Lowe River (33, fig. 7). A little gold, probably considerably less than 500 ounces, has been mined from a beach placer at the southwest end of Middleton Island (34, fig. 7), where wave action concentrates the heavy minerals from Tertiary marine clastic rocks deposited in part from floating ice. A few small nuggets, the largest worth 83 cents (gold at $20.67 per ounce), have been reported from this site.

YAKATAGA DISTRICT

The Yakataga district (fig. 7) includes the area drained by the Martin River and its tributaries and by streams flowing into the

Gulf of Alaska between the Copper River and long 141° W.

Since the 1890's a little gold has been recovered during most years from a beach about 18 miles long (36, fig. 7) near Yakataga. Total production probably has been about 15 to 16 thousand ounces. Most operations involve one or two men working with simple equipment. Storm waves are eroding a coastal plain composed of glacial and glaciofluvial material that rests on Tertiary marine rocks (Miller, 1957). The coastal plain sediments contain a little fine gold that wave action concentrates in patches, some of which are rich enough to be minable. The commonest constituent of the concentrates is garnet, found in association with magnetite, zircon, small amounts of chromite, rutile, gold, native copper, and various rock-forming minerals. No platinum has been identified in concentrate samples, although it occurs in similar beaches farther east at Lituya Bay in the Southeastern Alaska region. Gold was mined for a few years from bench and stream gravels of the White River (35, fig. 7), a proglacial stream that flows through material similar to that behind the Yakataga beach. The White River probably was a major contributor of sediments to the coastal plain.

KENAI PENINSULA REGION

The Kenai Peninsula region (pl. 1, fig. 10) is the Kenai Peninsula south of Turnagain Arm and west of the divide between Cook Inlet and Prince William Sound. It comprises the Homer, Hope, and Seward districts.

West of a line extending from the head of Kachemak Bay to Turnagain Arm near the mouth of the Chickaloon River, most of this region is considerably less than 1,000 feet above sea level, though rolling hills and a few steep-sided ridges rise to elevations of nearly 3,000 feet. The Kenai Mountains to the east are characterized by high relief, and many of the summits are between 4,000 and 6,000 feet in altitude. Deep fiords, many with glaciers at their heads, embay the coastline. Remnants of Pleistocene ice that covered the entire peninsula and extended far into the sea are preserved as alpine glaciers and as the Harding and Sargent Icefields. Proglacial lakes occupied much of the lowland during parts of Pleistocene time. Two large lakes in the lowlands, Skilak and Tustumena, lie behind recessional moraines, although the Kasilof River, which drains Tustumena Lake, has cut down to bedrock. The drainage of the northern part of the lowland is still not fully integrated. The entire region is free of permafrost.

The Kenai Mountains, the highest parts of which are virtually unexplored, are made up of limestone, chert, and tuff of Triassic

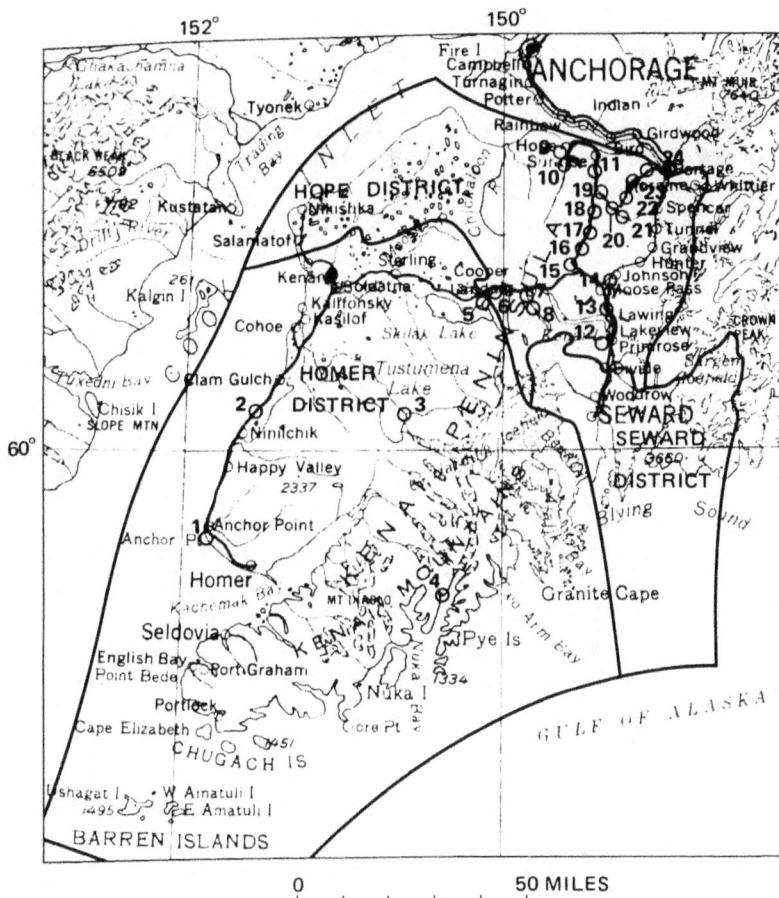

FIGURE 10.—Placer deposits in the Kenai Peninsula region.

Homer district

1. Anchor Point: Martin, Johnson, and Grant (1915, p. 110–111).
2. Ninilchik: Martin, Johnson, and Grant (1915, p. 111).
3. Indian Creek: Martin, Johnson, and Grant (1915, p. 111).
4. Morris, Sheridan, Kuppler and Lee: Martin, Johnson, and Grant (1915, p. 229).
5. Kenai River: Martin, Johnson, and Grant (1915, p. 111, 182, 197).

Hope district

6. Unnamed occurrences: Jasper (1967a, p. 42–43).
7. Cooper Creek: Martin, Johnson, and Grant (1915 p. 199–201). Kenai River: Martin, Johnson, and Grant (1915, p. 111, 181–182, 197–199).

Hope district—Continued

8. Unnamed occurrence: Jasper (1967a, p. 45).
9. Bear Creek: Moffit (1906a, p. 36). Resurrection Creek: Martin, Johnson, and Grant (1915, p. 193–195).
10. Palmer Creek: Moffit (1906a, p. 35–36). Resurrection Creek: Martin, Johnson, and Grant (1915, p. 193–195).
11. Sixmile Creek: Tuck (1933, p. 521, 526).
12. Porcupine (Primrose) Creek: Jasper (1967a, p. 42).
13. Falls Creek: Martin, Johnson, and Grant (1915. p. 202).
14. Upper Trail Lake: Jasper (1967a, p. 41).
15. Quartz Creek: Martin, Johnson, and Grant (1915, p. 201–202).
16. Unnamed occurrence: Jasper (1967a, p. 38).

age that rest on metamorphosed older volcanic and clastic rocks and are overlain by Jurassic volcanics and a thick sequence of intensely deformed, but only slightly metamorphosed, slate and graywacke, mainly of Late Cretaceous age (Kelly, 1963, p. 280–284; Berg and Cobb, 1967, p. 76). These rocks were intruded by Tertiary (?) dikes, sills, and stocks that range in composition from granite to peridotite (Berg and Cobb, 1967, p. 76; Richter, 1970, p. B4–B5). The lowland and adjacent parts of Cook Inlet are underlain by many thousands of feet of poorly consolidated, mainly continental, rocks of Tertiary age that rest on a basement of rocks similar to those exposed in the Kenai Mountains (Mac-Neil and others, 1961; Kelly, 1963). The Tertiary rocks are buried by Quaternary deposits except along sea cliffs around the southern part of the Kenai Lowland, in isolated inland exposures, and in a few small remnants resting on older rocks on the southeast shore of Kachemak Bay and at Port Graham.

Only gold, alloyed with silver, and chromite have been mined from lodes in the Kenai Peninsula region, although copper, lead, zinc, molybdenum, antimony, and nickel minerals have been found (Berg and Cobb, 1967, p. 73–82, fig. 14; Richter, 1970). The chromite is in two dunite and pyroxenite stocks in the southern part of the Homer district. Quartz veins in graywacke and slate and in small quartz diorite stocks and granite dikes carry gold and various sulfide minerals. The lode gold production of the region probably was about 19,000 ounces.

Placer gold was first discovered in Alaska on the Kenai River in 1848 (between loc. 5 and 7, fig. 10) by P. P. Doroshin, a mining engineer employed by the Russian-American Co. In 1850–51, he attempted to mine gold on a stream that flows into Skilak Lake and on two small tributaries of the Kenai River between Skilak

---

FIGURE 10.—Continued

Hope district—Continued

17. Canyon Creek: Moffit (1906a, p. 37–38). Juneau Creek: Martin, Johnson, and Grant (1915, p. 204–205). Mills Creek: Martin, Johnson, and Grant (1915, p. 204–205), Tuck (1933, p. 521–522).

18. Canyon Creek: Moffit (1906a, p. 37–38). Unnamed occurrence: Jasper (1967a, p. 37).

19. Gulch Creek: Martin, Johnson and Grant (1915, p. 206–207). Sixmile Creek, East Fork: Martin, Johnson, and Grant (1915, p. 265).

20. Silvertip Creek: Martin, Johnson, and Grant (1915, p. 206), Jasper (1967a, p. 36).

Hope district—Continued

21. Center Creek: Jasper (1967a, p. 36). Lynx Creek: Martin, Johnson, and Grant (1915, p. 207–208).

22. Bertha Creek: Jasper (1967a, p. 35). Granite Creek: Moffit (1906a, p. 40). Unnamed occurrence: Jasper (1967a, p. 35).

23. Ingram Creek, Granite Creek, and unnamed occurrence: Jasper (1967a, p. 34–35).

24. Ingram Creek: Jasper (1967a, p. 33).

Seward district

No placer occurrences.

and Kenai Lakes but failed to find enough to repay his efforts
(Moffit, 1906a, p. 8). Later placer mining was concentrated in
the parts of the Hope district where lode deposits were extensively explored and mined. A few streams and beaches in other
parts of the Kenai Peninsula region were worked on a small scale.
In the area around Nuka Bay, however, where there are many
gold-bearing lodes, placer gold has not been found. As production
statistics have generally included the output of Crow Creek and
neighboring streams in the Anchorage district in that credited
to the Kenai Peninsula region, accurate figures are not available.
The total for the Kenai Peninsula from about 1895, the first year
production was officially reported, through 1960 was probably
between 100 and 105 thousand fine ounces of gold and an unknown
amount of alloyed silver. Small-scale placer operations were reported in 1961 and 1962.

HOMER DISTRICT

The Homer district (fig. 10) is the area drained by the Kenai
River below and including Skilak Lake, by streams flowing into
the Gulf of Alaska from Callisto Head (the promontory east of
Bear Glacier) to the western end of the Kenai Peninsula, and by
streams flowing into Cook Inlet from the western end of the
Kenai Peninsula to the Kenai River. It also includes the Chugach
and Barren Islands.

Most of the small amount of placer gold mined in the Homer
district came from beach placers at Anchor Point (1, fig. 10),
north of Ninilchik (2, fig. 10), and possibly at other places along
the east shore of Cook Inlet. The most extensive operations were
at Anchor Point, where a ditch 2 miles long brought water for
sluicing fine gold from a thin layer of beach gravel about 2 feet
below the surface. Mining was carried on intermittently from
1889 until as recently as 1911. All the beach gold was derived
from glaciofluvial deposits. Rumors of the presence of platinum
at Anchor Point have not been confirmed.

Most of the bars and gravel terraces along the Kenai River
contain a little fine gold, as do the gravels of other streams in the
vicinity, but with the possible exception of a place on the Kenai
River near the head of Skilak Lake (5, fig. 10), there was no successful mining in the part of the Kenai River basin in the Homer
district. For several years a little small-scale mining was carried
on at Indian Creek (3, fig. 10), but an attempt to establish an
elaborate hydraulic operation there in 1903 failed. The gold at
Indian Creek and along the Kenai River was derived from thick

glacial and glaciofluvial deposits that had their source in the Kenai Mountains.

In the early 1900's, Morris, Sheridan, Kuppler, and Lee (4, fig. 10) staked a claim on the flat in front of McCarty Glacier near Nuka Bay, where they found large pieces of vein quartz carrying chalcopyrite. The bedrock source of this float material was not discovered.

HOPE DISTRICT

The Hope district (fig. 10) comprises the area drained by streams flowing into Cook Inlet from a point midway between Kenai and Salamatof to Portage and the area drained by the Kenai River above Skilak Lake. The eastern boundary is the divide between Cook Inlet and Prince William Sound.

Gold in many of the placer deposits of the Hope district was derived directly from nearby auriferous quartz veins, many of which were sites of lode mining. Some productive placers were in stream and bench gravels in which at least a portion of the heavy minerals was reconcentrated from glacial and glaciofluvial deposits derived from areas in which gold-bearing lodes are known or might reasonably be.

Although gold was discovered on the Kenai River in 1848, large-scale placer mining in the Hope district did not begin until nearly 50 years later. By 1900, most of the streams that ever became productive were being mined. Mills (17, fig. 10) and Canyon (17, 18, fig. 10) Creeks were the most productive streams in the district. On both creeks, bench gravels as well as the present stream beds were mined, mainly by hydraulic methods. Mills Creek was worked as recently as 1961, the last year in which a placer operation larger than a small drift mine was reported from any part of the Kenai Peninsula. Small dredges were used on the Kenai River (7, fig. 10) and on Sixmile (11, fig. 10) and Resurrection (9, 10, fig. 10) Creeks, but the results evidently were not satisfactory, as other mining methods supplanted the dredges after a year or two of operation on Sixmile and Resurrection Creeks. All mining on the Kenai River ceased with the abandonment of the dredge in 1913.

Few data are available on the minerals, other than gold, in concentrates from placers in the Hope district. Native silver was reported from Bear Creek (9, fig. 10), native copper from Lynx Creek (21, fig. 10), and a few sulfide minerals from other streams. In 1966 Jasper (1967a) collected concentrate samples along highways on the Kenai Peninsula as part of a regional geochemical study and reported scheelite, gold, or various sulfide minerals from many localities. Of particular interest was cinnabar from Bertha Creek (22, fig. 10) and from a gully (8, fig. 10) near Cooper Creek, as no lode occurrences of mercury have been reported within 50 miles.

SEWARD DISTRICT

The Seward district (fig. 10) is the area drained by streams flowing into Resurrection Bay and Blying Sound from Callisto Head on the west to Cape Junken on the east.

Although a few nonproductive gold lodes and several occurrences of copper minerals have been reported, no placer deposits have been found in the district.

KODIAK REGION

The Kodiak region (pl. 1, fig. 11) includes Kodiak, Afognak, and the Trinity Islands and nearby small islands. The region, classified as a single district, is characterized by mountains with summits 2,000–4,000 feet in altitude and by gently rolling uplands. Long narrow inlets extend well into the interiors of Kodiak and Afognak Islands.

Most of the region is underlain by Cretaceous graywacke, slate, and conglomerate that rest on older Mesozoic marine and volcanic rocks containing a few small mafic and ultramafic bodies and by Tertiary quartz diorite plutons, some of batholithic dimensions. Younger Tertiary marine and continental rocks form the Trinity Islands and a fringe along the southeastern coast of Kodiak Island. Quaternary glacial and fluvial deposits mantle bedrock in low areas on western Kodiak Island, at the heads of bays, and along some of the larger streams. Long faults extend the length of the major islands, giving the region a pronounced northeast-trending grain that, if prolonged, would join generally similar features at the southwestern end of the Kenai Peninsula. The foregoing summary is based on a recent map by Moore (1967) and an earlier report by Capps (1937).

The Kodiak region was covered by Pleistocene ice that extended from the crest of the Aleutian Range across the islands and several tens of miles into the Pacific Ocean. The ice removed most unconsolidated material and any placer deposits that may have been formed in preglacial valleys. Ice remains in a few cirque glaciers on the highest peaks on Kodiak Island.

An unknown, but probably small, amount of gold was mined from several lode deposits in the Kodiak region, mainly before World War I and about 1935. Lode occurrences of tungsten and

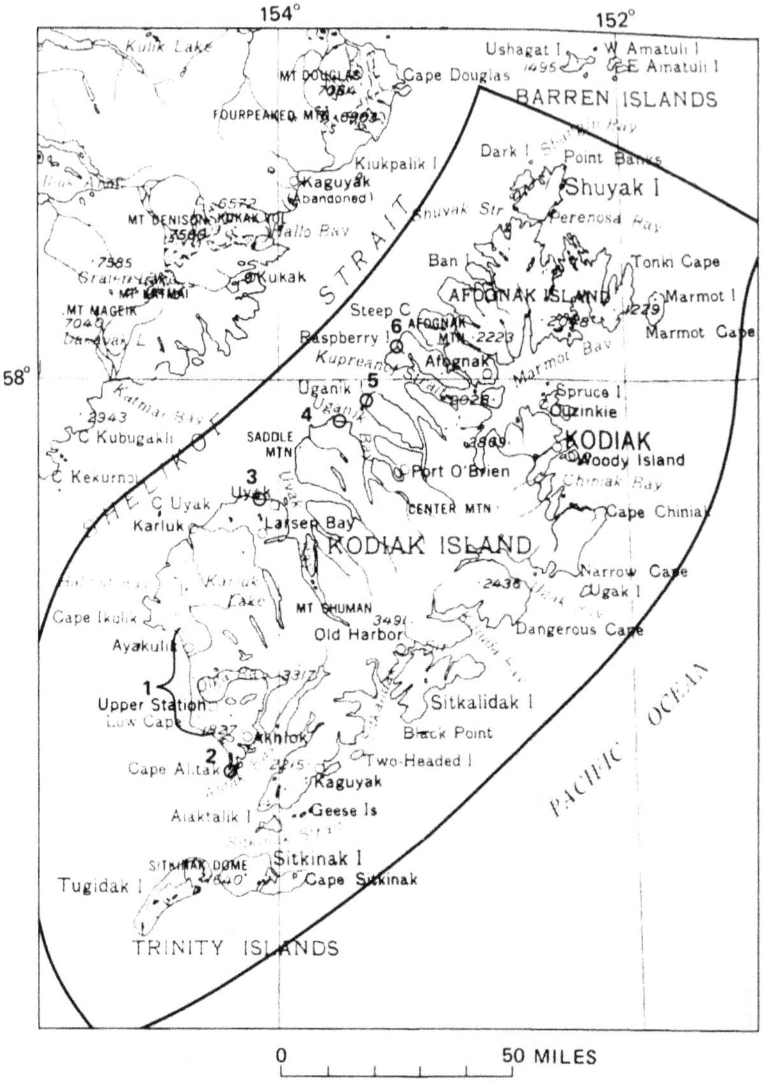

FIGURE 11.—Placer deposits in the Kodiak region.

1. Kodiak Island, west coast beaches (Ayakulik River, Red River, Canvas Point): Maddren (1919, p. 311–319).
2. Cape Alitak: Capps (1937, p. 172–173).
3. Sevenmile Beach: Martin (1913, p. 134).
4. Miners Point Beach: Capps (1937, p. 172).
5. Uganik Beach: Capps (1937, p. 172).
6. Raspberry Beach (Driver Bay): Capps (1937, p. 172).

copper proved to be too small and of too low grade to be mined (Berg and Cobb, 1967, p. 82–88, fig. 15).

The only placers that have been found in the region are in beach deposits, where gold was concentrated from lean glacial outwash and till. Mining was on a small scale, using rockers and portable sluice boxes operated with water brought from nearby lakes by ditches and canvas hose. Most of the activity was on the beaches along the west coast of Kodiak Island (1, fig. 11), where wave action concentrated heavy minerals in a veneer of material in transit across a planation surface cut on glacial deposits. By far the greatest part (95 percent) of the concentrates was magnetite. Other heavy minerals include pyrite, chromite, gold, and a little platinum. Ultramafic bodies that had been overridden by ice were the original sources of the chromite and platinum and at least some of the magnetite. The placer at Cape Alitak (2, fig. 11) is unusual in that the small amount of fine gold recovered came from dune sands derived, either directly or by way of beach deposits, from bluffs of glacial material.

The total production of gold from beaches in the Kodiak region is not known, as records combine data from all of southwestern Alaska, but it probably was not more than a few thousand ounces at most. The only beach mining reported since World War II was in 1951–52, when two men were working on the west coast of Kodiak Island.

KUSKOKWIM RIVER REGION

The Kuskokwim River region (pl. 1, figs. 12, 13, 16) includes Nunivak and Nelson Islands and the mainland area drained by streams flowing into Baird Inlet, Etolin Strait, and Kuskokwim Bay. It comprises the Aniak, Bethel, Goodnews Bay, and McGrath districts.

The Kuskokwim River region is dominated by the Kuskokwim Mountains, a succession of rounded northeast-trending ridges 1,500–2,000 feet in altitude surmounted locally by rugged mountains as much as 2,000 feet higher. Other upland areas include the Kilbuck Mountains and parts of Nunivak Island. The eastern boundary of the region is the crest of the southern Alaska Range, most of which is more than 6,000 feet in elevation; the highest peak is Mount Foraker (17,395 ft). About a third of the region consists of lowlands less than 1,000 feet in altitude along the major rivers.

The following summary of the geology of the region is based mainly on reports by Cady and others (1955), Hoare (1961), Hoare and others (1968), and Reed and Elliott (1970), and on informal discussions with William H. Condon, Joseph M. Hoare, and Bruce L. Reed.

The oldest rocks, a narrow belt of gneiss and schist about 75 miles long in the western part of the region, may be Precambrian in age. Paleozoic sedimentary rocks range in age from Cambrian to Devonian in the Alaska Range and from Devonian to Permian in the Kuskokwim Mountains. A great mass, possibly as much as 5 or 6 miles thick, of Carboniferous, Mesozoic, and Tertiary graywacke, shale, conglomerate, volcanic rocks, and limestone underlies most of the Kuskokwim River region west of the Alaska Range and its foothills. These rocks were displaced by major northeast-trending zones of strike-slip faulting, some of which can be traced far beyond the boundaries of the region (Grantz, 1966).

Upper Cretaceous and Tertiary plutons, dikes, and sills that range in composition from ultramafic to felsic intruded the older rocks in nearly all parts of the region. Tertiary and Quaternary basaltic lava flows and associated tuffs cover most of Nunivak and Nelson Islands.

Quaternary fluvial and glacial deposits that locally have been reworked by wave and wind action floor the lowlands and valleys. Most of the surficial deposits are permanently frozen except near large bodies of water; many of the mountainous parts of the region are in areas underlain by discontinuous or isolated masses of permafrost. Only the Alaska Range, the mountains in the southwestern part of the region, and isolated summits of the Kuskokwim Mountains have nourished Pleistocene glaciers, remnants of which are preserved in cirques and valleys in the Alaska Range.

Lodes in the Kuskokwim River region have been the source of most of the mercury mined in Alaska; a total of somewhat more than 35,000 76-pound flasks was produced between 1902 and 1967 (Alaska Division of Mines and Minerals, 1967, p. 8). Some gold and a little antimony ore as a byproduct have also been mined. Other lodes, some fairly extensively prospected, contain gold and various copper, lead, zinc, molybdenum, tungsten, bismuth, antimony, mercury, manganese, and uranium minerals (Berg and Cobb, 1967, p. 88–97, figs. 16–18). Reed and Elliott (1968a, 1968b, 1970) described many occurrences of base and precious metals in the eastern parts of the Aniak and McGrath districts. Some are in bedrock, others consist of mineralized float in and near contact zones around granitic plutons.

Gold lodes north of Medfra (near loc. 5 and 6, fig. 16) were the source of 40 to 60 thousand ounces of gold and a little silver.

Lode cinnabar was discovered by the Russians in the Kuskokwim River region about 1838 and prospectors looking for gold passed through the region as early as 1889, but no placer deposits were found until about 1900, when a number of men from Nome participated in a stampede set off by vague rumors of a discovery on a stream called "Yellow River," said to be somewhere in the Kuskokwim Valley (Maddren, 1915, p. 299–300). From 1908 through 1960 about 650,000 fine ounces of gold (3.2 percent of the total Alaskan placer-gold production) was recovered from placers in the region. Mining has been reported in every year since 1960, but production data have not been made public. More than half a million ounces of platinum-group metals have been recovered from placers in the Goodnews Bay district (Mertie, 1969, p. 77, 79, 87). Small amounts of cinnabar and scheelite were mined from streams draining lodes that carry these minerals.

ANIAK DISTRICT

The Aniak district (fig. 12) is the area drained by the Kuskokwim River and its tributaries above Bethel as far as (and including) the Stony River.

The principal center of mining in the district was near Nyac, where placer deposits on the Tuluksak River (6–8, fig. 12) and Bear (9, 10, fig. 12) and California (8, fig. 12) Creeks and their tributaries were mined from 1909 until the end of the 1964 season when the last of three dredges was shut down. The source of most of the gold was low-grade gold- and sulfide-bearing quartz veins in contact zones between Cretaceous volcanic rocks and Tertiary granitic plutons and in the plutons themselves. Cinnabar, probably originally assiciated with extensively altered diabasic dikes or sills, constituted a large part of the concentrates of a dredge operating on Bear Creek near the mouth of Bonanza Creek (10, fig. 12). A little platinum was recovered by some of the dredges in the area, but the amount is unknown. Less rich and less extensive placer deposits similar to those near Nyac were prospected or mined on Granite (5, fig. 12), Ophir (12, fig. 12), and other creeks.

Successful mining has been carried on for many years on Marvel Creek (15, fig. 12) and, to a lesser extent, on other tributaries of the Salmon River. The gold in the placers was derived from contact zones between granitic plutons and clastic Cretaceous rocks. Unlike the area near Nyac, sulfide minerals have not been reported in concentrates or in quartz veinlets in contact zones. Until 1966, when a small dredge was brought to Marvel Creek from Nyac, all operations were by nonfloat, hydraulic, and hand methods. Farther south, on Canyon Creek (1, fig. 12), one of the headwaters of the Kwethluk River, there has also been mining in practically every year since gold was discovered there in 1913. Canyon Creek crosses a contact zone, the probable source of the gold in the placers, between a quartz porphyry body and Paleozoic or Mesozoic clastic rocks. Hoare and Coonrad (1959a) found evidence of prospecting or small-scale mining on Columbia (2, fig. 12) and Rocky (3, fig. 12) Creeks, but no reports of successful mining on these or other streams in the basin of the Kisaralik River have been published. A report of cassiterite from the Riglagalik (Martin, 1919, p. 20) River, another name for the Kisaralik, was probably false.

After the Nyac area, the most productive part of the Aniak district was the basin of Crooked Creek (21, fig. 12), where benches about 1 mile wide lying east and southeast of and parallel to the main stream and its principal tributary, Donlin Creek (23, 24, fig. 12), were mined from about 1910 until at least as recently as 1956. The gold was derived from small quartz fracture fillings in Cretaceous graywacke and shale near small silicified porphyritic albite rhyolite intrusive bodies. The richest placers were on Snow and other left-limit gulches (22, fig. 12) in which gold from the benches had been further concentrated. In addition to gold, concentrates contained magnetite, garnet, scheelite, cassiterite, pyrite, cinnabar, and stibnite. Only the gold was saved. Julian Creek (26, fig. 12), a tributary of the George River, drains a source area similar to that of the Crooked Creek-Donlin Creek placers and was mined sporadically from about 1911 to 1939. Concentrate samples contained gold, pyrite, some cinnabar, and traces of monazite. Other streams, such as Murray Gulch (20, fig. 12) near Napamute and small tributaries of the Kuskokwim River between Crooked Creek and Sleetmute (25, 27–29, fig. 12), drain geologically similar areas and were sites of prospecting or mining before World War II, but only Murray Gulch and New York Creek (20, fig. 12) have ever been listed as producing streams.

In the upper Holitna River basin, gold has been recovered from Taylor (31, fig. 12) and Fortyseven (19, fig. 12) Creeks. Cassiterite, cinnabar, and pyrite accompany the gold in Taylor Creek; all were probably derived from mineralized zones in Cretaceous clastic rocks in the Taylor Mountains, where they were altered to hornfels around a quartz monzonite stock, or from mineralization associated with small albite rhyolite intrusive bodies in the nearby Little Taylor Mountains. Somewhat more than 2,000 ounces of gold is said to have been recovered from Taylor Creek. Fortyseven Creek drains a ridge on which there is a scheelite- and gold-bearing lode in a silicified shear zone in graywacke and shale. Both gold and scheelite have been recovered from placer deposits below the lode. Scheelite has also been reported in an indefinitely described area west of the Horn Mountains about 15

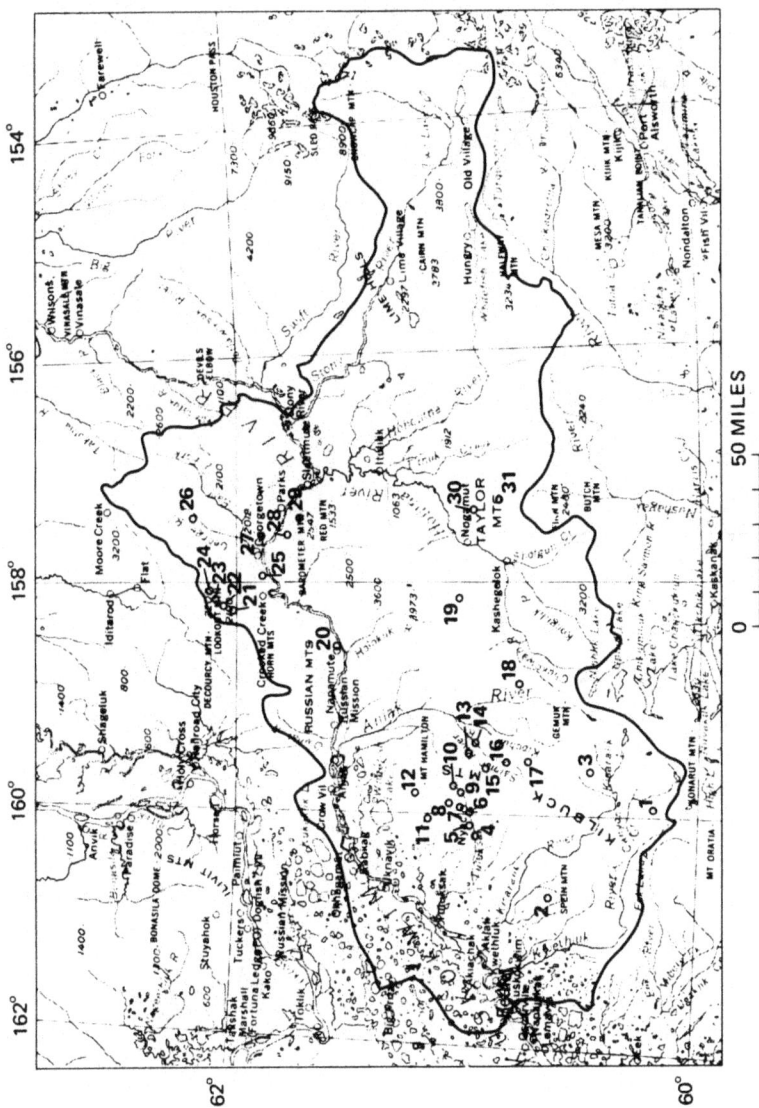

FIGURE 12.—Placer deposits in the Aniak district.

FIGURE 12.—Continued.

1. Canyon Creek: Maddren (1915, p. 356-357), Alaska Division of Mines and Geology (1968, p. 51).
2. Columbia Creek: Hoare and Coonrad (1959a).
3. Rocky Creek: Hoare and Coonrad (1959a).
4. Tuluksak River: Maddren (1915, p. 332).
5. Granite Creek: Maddren (1915, p. 331), Hoare and Coonrad (1959a).
6. Tuluksak River: Hoare and Coonrad (1959a).
7. Tiny Gulch: Maddren (1915, p. 328-329). Tuluksak River: Maddren (1915, p. 331), Alaska Division of Mines and Minerals (1960, p. 77).
8. California Creek: Hoare and Coonrad (1959b). Rocky (Rock) Creek: Alaska Division of Mines and Minerals (1960, p. 77). Tuluksak River: Maddren (1915, p. 331), Alaska Division of Mines and Minerals (1960, p. 77).
9. Bear Creek: Maddren (1915, p. 309-321, 328-330). Hoare and Coonrad (1959b), Mertie (1969, p. 89-90). Spruce Creek: Maddren (1915, p. 311-312, 321-322).
10. Bear Creek: Maddren (1915, p. 309-321, 328-330). Hoare and Coonrad (1959b), Mertie (1969, p. 89-90). Bonanza Creek: Maddren (1915, p. 311-312, 327-329).
11. Bogus Creek: Maddren (1915, p. 331-332).
12. Ophir Creek: Maddren (1915, p. 332-336).
13-14. Dominion Creek: Maddren (1915, p. 336-338), Hoare and Coonrad (1959a).
15. Marvel Creek: Maddren (1915, p. 339-346), Alaska Division of Mines and Minerals (1966, p. 10).
16. Fisher Creek: Maddren (1915, p. 346-347).
17. Cripple Creek: Maddren (1915, p. 347-351), Hoare and Coonrad (1959a).
18. Cinnabar Creek: Sainsbury and MacKevett (1965, p. 42-43).
19. Fortyseven Creek: Cady and others (1955, p. 119-121).
20. Murray Gulch: Maddren (1915, p. 353-355). New York Creek: Maddren (1915, p. 353).
21. Crooked Creek: Maddren (1915, p. 351-353), Cady, and others (1955, p. 118).
22. Lewis, Quartz, and Queen Gulches: Cady and others (1955, p. 118). Ruby Gulch: Maddren (1915, p. 352-353), Cady and others (1955, p. 118). Snow Gulch: Cady and others (1955, p. 118).
23-24. Donlin Creek: Cady and others (1955, p. 68-69, 116, 118).
25. Central Creek: Cady and others (1955, p. 120).
26. Julian Creek: White and Killeen (1953, p. 16, 18), Cady and others (1955, p. 71, 116, 119).
27. California Creek: Cady and others (1955, p. 120).
28. Eightmile Creek: Cady and others (1955, p. 120).
29. Fuller Creek: Cady and others (1955, p. 120).
30. Stevens Creek: Cady and others (1955, p. 121).
31. Taylor Creek: Cady and others (1955, p. 71, 116, 119).

miles north of Napamute and wolframite has been reported in float on a ridge west of Stevens Creek (30, fig. 12).

Placer cinnabar was mined from Cinnabar Creek (18, fig. 12), immediately downstream from the Cinnabar Creek lode mine, described by Sainsbury and MacKevett (1965, p. 38–40). Although the placer production probably was small, it is of particular interest because it is one of the few examples of successful primary production of a nonprecious metal from an Alaskan placer mine.

BETHEL DISTRICT

The Bethel district (fig. 13) includes the area drained by the Kuskokwim River below Bethel and by streams flowing into Baird Inlet, Etolin Strait, and Kuskokwim Bay as far south as, but

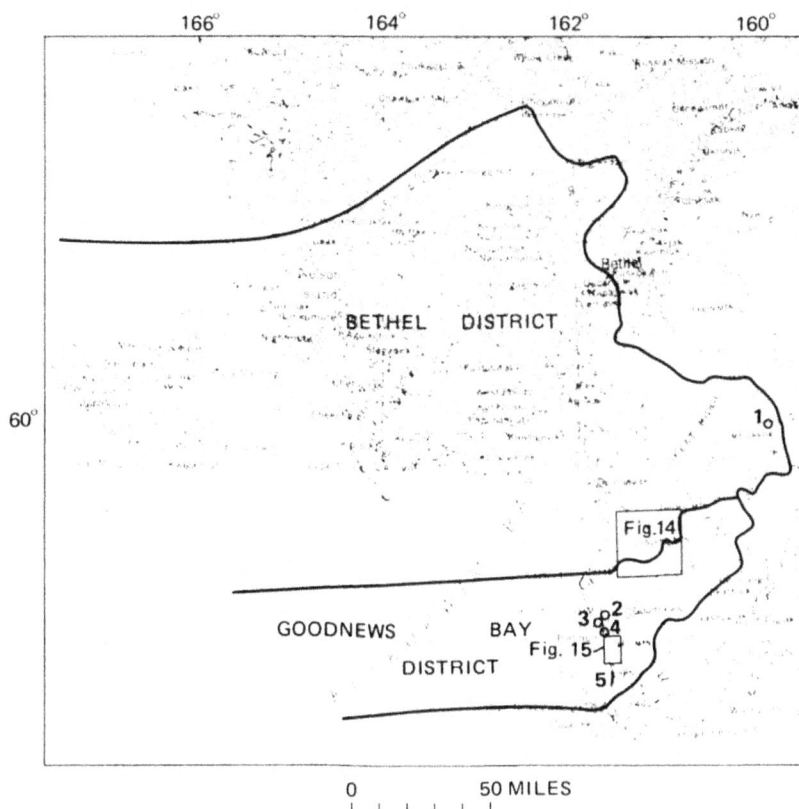

FIGURE 13.—Placer deposits in the Bethel and Goodnews Bay districts.

Bethel district
1. Rainy Creek: Rutledge (1948, p. 3, 7).
Goodnews Bay district
2–3. Goodnews Bay: Berryhill (1963, p. 13–16).

Goodnews Bay district—Continued
4. Goodnews Bay: Berryhill (1963, p. 13–15).
5. Chagvan Bay: Berryhill (1963, p. 13–16).

excluding, Carter Bay. It also includes Nelson and Nunivak Islands.

Practically all of the placer mining in the district has been on tributaries of the Arolik River (fig. 14). No lode source for the gold and platinum in the placers has been found in this area, but the distribution of the deposits suggests that much of the gold may have been derived from contact zones around small granitic

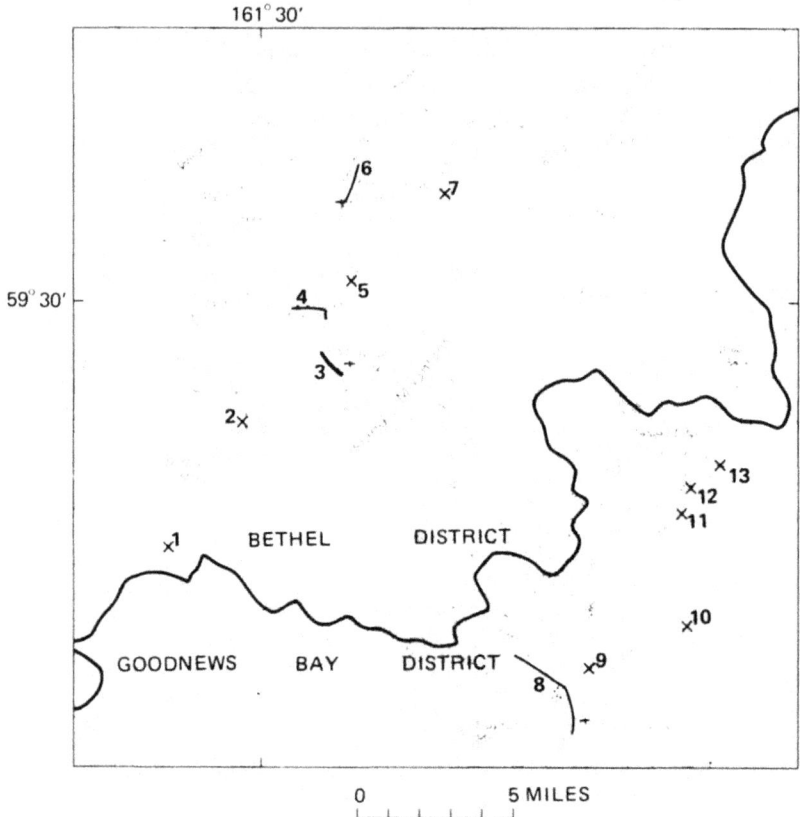

FIGURE 14.—Placer deposits in the Arolik River-Goodnews River area.

Bethel district

1. Jacksmith Creek tributary: Unpub. data.

2. Domingo Creek: Hoare and Coonrad (1961).

3. Kowkow Creek: Harrington (1921, p. 227), Smith (1930b, p. 52–53).

4. Butte Creek: Harrington (1921, p. 227), Mertie (1969, p. 89–90).

5. Fox Creek: Unpub. data.

6. Snow Gulch: Smith (1942b, p. 54), Mertie (1969, p. 89–90).

7. Tyone Creek: Hoare and Coonrad (1961).

Goodnews Bay district

8. Slate Creek: Smith (1939b, p. 61). Wattamuse Creek: Harrington (1921, p. 225–226), Smith (1942b, p. 54).

9. Olympic Creek: Smith (1933b, p. 44).

10. Fox Creek (Gulch): Hoare and Coonrad (1961).

11–12. Bear Creek: Harrington (1921, p. 226–227).

13. Canyon Creek: Hoare and Coonrad (1961).

plutons. Altered ultramafic bodies such as one exposed along the summit of Island Mountain might have contributed the platinum-group metals. The minable placer deposits are probably the result of reconcentration of heavy minerals from glaciofluvial and glacial deposits.

Gold was discovered' in the Arolik River basin in about 1900. At first only Butte Creek (4, fig. 14) was worked, but from 1913 until World War II, Kowkow Creek (3, fig. 14) and Snow Gulch (6, fig. 14) also were extensively mined. A little platinum was recovered with the gold from all these creeks. Signs of old placer activity have been reported on other creeks in the neighborhood, but the names of the miners and the results of their work are not known.

The only other place in the Bethel district where there was successful placer mining is on Rainy Creek (1, fig. 13) ; some gold, probably reconcentrated from glaciofluvial or glacial deposits, and about 1 ton of cinnabar concentrates were recovered. The cinnabar was derived from a low-grade lode at the head of Arsenic Creek (Rutledge, 1948), a small tributary of Rainy Creek. In 1914, a little coarse gold was groundsluiced from a claim somewhere on Kapon Creek, a headwater tributary of the Eek River a few miles north of Rainy Creek (Maddren, 1915, p. 357). Associated with the gold were magnetite sand, small cinnabar pebbles, and arseno-pyrite.

GOODNEWS BAY DISTRICT

The Goodnews Bay district (fig. 13) is the area drained by streams flowing to Kuskokwim Bay from (and including) Carter Bay to Cape Newenham.

The first mining in the district was on tributaries of the Goodnews River that head in the divide separating the basins of the Goodnews and Arolik Rivers (fig. 14). Claims were staked on Bear Creek (11–12, fig. 14) in 1916, and soon after that richer deposits were found on Wattamuse Creek (8, fig. 14). Mining was reported for nearly every year thereafter on Wattamuse Creek or on Slate Creek (8, fig. 14), into which Wattamuse flows, until 1961. The bedrock source of the gold is probably quartz veins in

FIGURE 15 (right).—Placer deposits in the Salmon River area.

1. Platinum-Salmon River beach: Berryhill (1963, p. 13–16).

2. Red Mountain: Mertie (1969, p. 80–81).

3–8. Red Mountain: Berryhill (1963, p. 13–16).

9. Red Mountain: Mertie (1969, p. 84).

10. Clara Creek: Mertie (1940, p. 57–59, 67–70), Mertie (1969, p. 77, 85–86).

11. Dowry Creek: Mertie (1969, p. 77, 85–86).

12. Dry Gulch: Mertie (1940, p. 57–58). Fox Gulch: Mertie (1940, p. 57–64), Mertie (1969, p. 80–81, 85–86). Platinum Creek: Mertie (1940, p. 57–64), Mertie (1969, p. 80–81, 84–87). Salmon River: Mertie (1969, p. 77, 79–88). Squirrel Creek: Mertie (1940, p. 57–58, 64–65).

contact zones around one or more granitic plutons in the divide in which all these streams head.

The most extensive mining in the Goodnews Bay district is on the Salmon River and its tributaries (fig. 15), where platinum

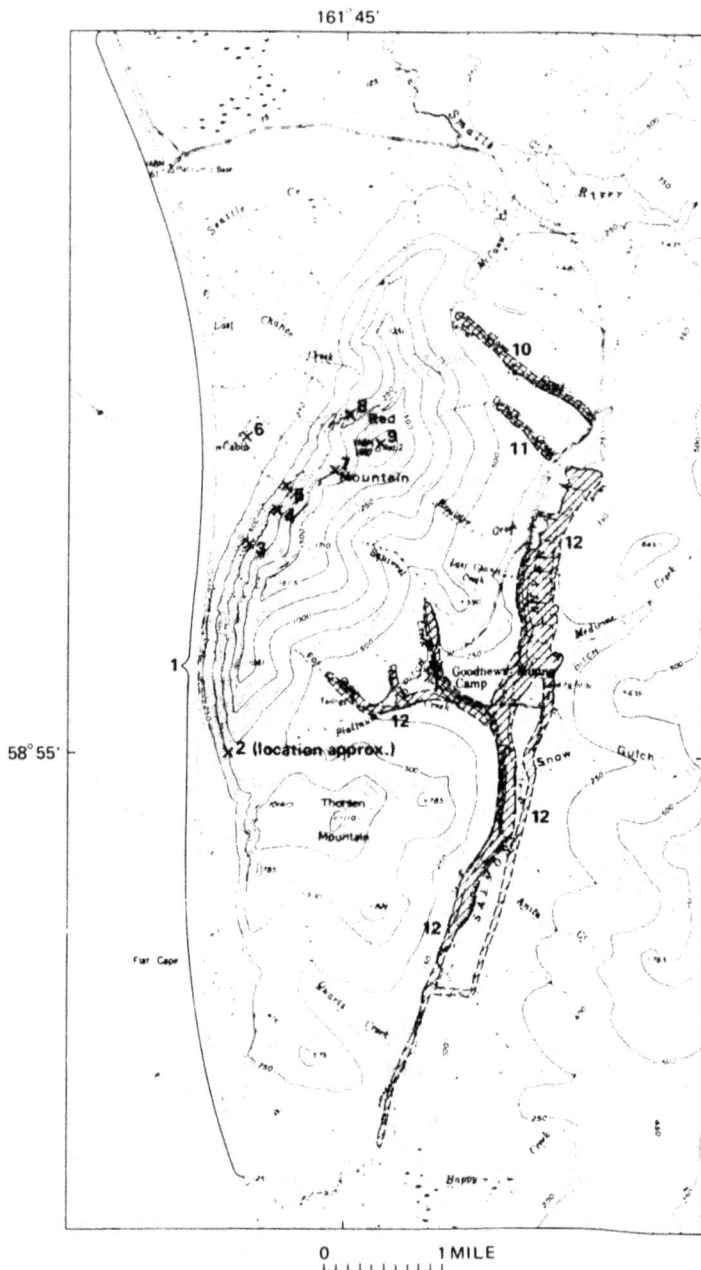

was discovered at the mouth of Fox Gulch (12, fig. 15) in 1926. The Clara Creek Mining Co. worked out Clara (10, fig. 15) and Dowry (11, fig. 15) Creeks between 1936 and 1940; and the Goodnews Bay Mining Co., now the sole operator in the area, began mining on Platinum Creek (12, fig. 15) in 1934. Mining has been with mechanized equipment, including a dredge that has worked or will work most of the gravel in the Salmon River valley (12, fig. 15). Total production from the Salmon River and its tributaries from 1934 to 1966 is estimated to have been well over half a million troy ounces of platinum-group metals (a major portion of the United States primary production) and a small amount of gold. Data on production since 1966 are not available, but operations have been on about the same scale as earlier years. The average percentages for platinum-group metals, gold, and impurities based on data from the Goodnews Bay Mining Co. (Mertie, 1969, p. 87) are: platinum, 73.62; iridium, 9.94; osmium, 1.89; ruthenium, 0.15; rhodium, 1.15; palladium, 0.34; gold, 2.06; and impurities, 10.85. The source of the platinum-group metals is the ultramafic body, composed of dunite and serpentinite, that makes up Red Mountain. Although no minable platinum has been found in any part of Red Mountain, a concentrate of residual material from near the top of the mountain (9, fig. 15) contained platinum, as did material in a small amphitheater (2, fig. 15) west of the divide at the head of Platinum Creek. Chromite, which is a component of concentrates from the Salmon River and some of its tributaries, is a constituent of the ultramafic body. The gold in the placers was derived from glacial deposits that came from the east.

Exploratory drilling by the Goodnews Bay Mining Co. in areas where material derived from Red Mountain would logically be expected failed to discover any workable platinum deposits outside of the Salmon River valley, although one drill hole encountered platiniferous gravel north of Red Mountain in what was interpreted as a pothole in bedrock. The general absence of platinum is interpreted to be the result of removal of preglacial surficial material by Pleistocene ice. Any beach placers that might have formed west of Red Mountain during times of lower sea level would now be far offshore (Mertie, 1969, p. 80–81). A reconnaissance study of beach sands along part of Kuskokwim Bay (Berryhill, 1963) failed to find any platinum, although there was chromite in samples from Goodnews Bay (2–4, fig. 13), from the beach west of Red Mountain (1, fig. 15), and from the beach north of the entrance to Chagvan Bay (5, fig. 13). Private companies have recently investigated possible submarine placer deposits west and south of the mouth of the Salmon River, but the results of their investigations have not been made public (1970).

McGRATH DISTRICT

The McGrath district (fig. 16) is the area drained by the Kuskokwim River above the Stony River.

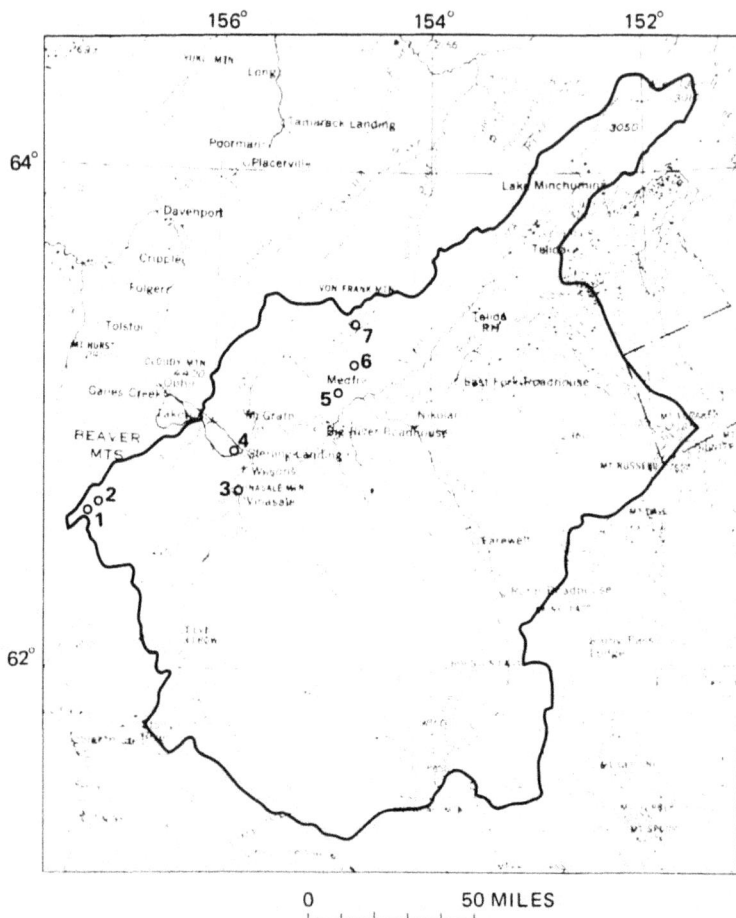

FIGURE 16.—Placer deposits in the McGrath district.

1. Moore Creek: Mertie (1936, p. 223-224), White and Killeen (1953, p. 16, 18).
2. Fourth of July Creek: Brooks (1912, p. 40), Smith (1939a, p. 59).
3. Alder Creek: Smith (1934b, p. 45), E. H. Cobb (this rept.)
4. Candle Creek: Mertie (1936, p. 197-198), White and Killeen (1953, p. 16, 18).
5. Crooked Creek: Brown (1926a, p. 139). Eagle Creek: White and Stevens (1953, p. 16, 18-19).
6. Birch Gulch: Mertie (1936, p. 195-196). Crystal Gulch: Brown (1926a, p. 138), Mertie (1936, p. 196-197). Hidden Creek: Brown (1926a, p. 136-137), Mertie (1936, p. 193-195). Holmes Gulch: Brown (1926a, p. 137), Mertie (1936, p. 196). Riddle Gulch: Mertie (1936, p. 194). Ruby Creek: Brown (1926a, p. 138), Mertie (1936, p. 196-197, White and Stevens (1953, p. 12, 15-16, 19).
7. Cottonwood Creek: Brown (1926a, p. 118, 141).

All the placer mining in the district has been on streams that cross contacts between small plutons, commonly near quartz monzonite in composition, and Paleozoic limestone or Cretaceous sandstone and shale. The sources of most of the heavy minerals in the placers are mineralized quartz veins in the contact zones and in the plutons, but only in the Nixon Fork and Eagle Creek areas (5, 6, fig. 16) were any of the lodes mined (Berg and Cobb, 1967, p. 96–97).

Gold was discovered on Moore Creek (1, fig. 16) in 1910 and was mined intermittently until at least as recently as 1967. Both stream and bench placers were worked, mainly by hydraulic methods. The principal minerals in the concentrates were cinnabar and chromite, neither of which was saved; accompaning minerals were zircon, magnetite, pyrite, and scheelite in smaller amounts. The chromite was probably derived from a mafic igneous rock, pebbles of which are in the creek gravels. There was a little mining and prospecting on Fourth of July Creek (2, fig. 16), which rises near the head of Moore Creek, where a small pluton intrudes Cretaceous clastic rocks.

The most productive stream in the McGrath district probably was Candle Creek (4, fig. 16), where gold was discovered in 1913 and was mined until World War II. The only dredge in the district operated there from 1917 until 1926. The gold, some of which was in nuggets weighing one or two ounces, was derived from quartz veins in quartz monzonite and Cretaceous sandstone and shale adjacent to the intrusive body. Cinnabar was so abundant in concentrates that for several years it was saved and retorted and the mercury produced was sold locally. Other heavy minerals in concentrates included magnetite, scheelite, and monazite(?). Alder Creek (3, fig. 16) is geologically similar to Candle Creek, but mining was on a small scale and was carried on for only a few years. Between 1929 and 1933 about 65 ounces of gold was recovered from a cut about 13,600 square feet in area. Large boulders hampered the hand-mining operation. No cinnabar was reported in the concentrates, which contained considerable scheelite and some magnetite, stibnite, and bismuth.

Placer gold was discovered on Hidden and Ruby Creeks (6, fig. 16) in 1917 and soon thereafter on neighboring streams, all of which drain an area in which a quartz monzonite stock had intruded Paleozoic limestone. By following placer deposits upstream, prospectors discovered lodes that yielded several tens of thousands of ounces of gold. The stream placers were mined on a small scale until the middle 1930's, but the total production from them was much less than that from the lodes. Native bismuth, a constituent of the lodes, was also common in placer concentrates. Other heavy minerals included magnetite, scheelite, cassiterite, hematite, ilmenite, sphene, zircon, and thorianite. Eagle Creek (5, fig. 16), a few miles to the southwest, is in a similar geologic setting and drains an area where there was a small lode-gold mine. The stream was mined on a small scale in the 1920's and early 1930's. Bismuth was not reported, but a concentrate sample contained allanite, garnet, ilmenite, scheelite, thorianite, and a trace of sphene. Colors of gold have been reported from other streams in this part of the McGrath district, but there is no record of successful mining on any of them.

NORTHERN ALASKA REGION

The northern Alaska region (pl. 1, fig. 17) is the part of Alaska drained by streams flowing into the Arctic Ocean and Chukchi Sea from the Alaska-Yukon boundary (long 141° W.) on the east to and including the Wulik River on the southwest. The region is divided into the Barrow, Canning, Colville, Lisburne, and Wainwright districts.

The major physiographic features are the Brooks Range, whose summits are between 4,000 and 8,000 feet in altitude along the southern boundary of the region; the northern foothills; and a lake-dotted coastal plain (less than 600 feet above sea level) that slopes gently to the Arctic Ocean.

Recent reports and maps by Brosgé and Tailleur (1970), Gates and Gryc (1963), Lathram (1965), and Tailleur (1969) summarize the geology of northern Alaska.

The bedded rocks form two east-trending belts: a southern belt of Paleozoic carbonate and clastic rocks and a northern belt of Mesozoic clastic and volcanic rocks. Tertiary conglomerate, sandstone, and siltstone underlie parts of the Canning and Colville districts. The Paleozoic and Mesozoic rocks were displaced along eastward-striking folds, nappes, and thrust faults, along some of which there were many miles of movement. In the Romanzof Mountains in the eastern part of the Canning district, granitic rocks invaded bedded rocks of pre-Mississippian age. In the western Brooks Range and the Arctic Foothills, Jurassic mafic intrusive bodies cut the older rocks. The Arctic Coastal Plain is covered by Quaternary silt, sand, and gravel.

In northern Alaska, only the DeLong Mountains and other parts of the Brooks Range supported extensive Pleistocene glaciers, small remnants of which remain in isolated cirques. The entire region is in a zone of continuous permafrost, which reaches thicknesses of many hundreds of feet, a condition that causes serious

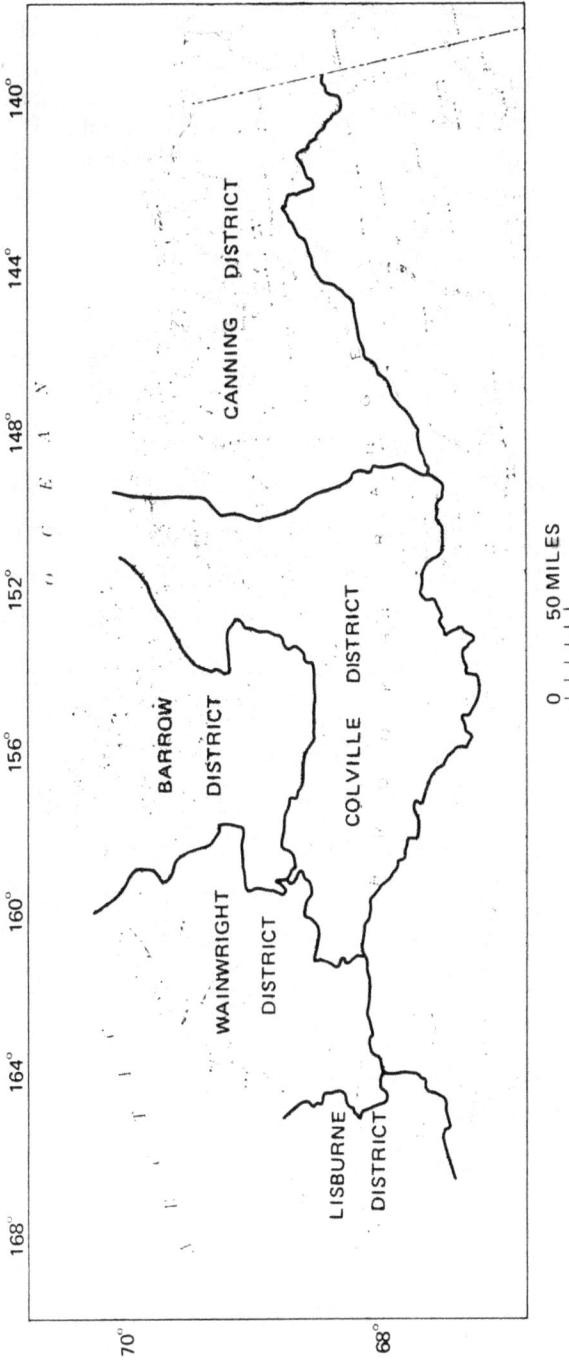

FIGURE 17.—Northern Alaska region and included districts.

engineering problems (Ferrians, Kachadoorian, and Green, 1969).

No metallic lode deposits have been reported in northern Alaska, although specimens of cinnabar purported to have come from the Canning River were sent to an assay office of the Alaska Territorial Department of Mines (now State of Alaska Division of Mines and Geology) some time before 1947 (Anderson, 1947, p. 34). In the Romanzof Mountains Paleozoic conglomerate contains cassiterite and tourmaline that may have been derived from veins in a contact zone around a granitic pluton (Reed, 1968, p. 33–34). Recent stream-sediment sampling in and near the Romanzof Mountains (Brosgé, Reiser, and Estlund, 1970) indicates possibly anomalous concentrations of some metals (not including gold or silver), particularly in streams that drain an area underlain by granitic rocks containing pyritic zones.

In the late 1880's, pyrite(?) said to contain $3.50 to $8.50 in gold per ton was reported on a tributary of the Pitmegea River in the northeastern part of the Lisburne district (Smith and Mertie, 1930, p. 339), an area in which geologic conditions are not favorable for the occurrence of auriferous lodes. From time to time Eskimos have brought specimens of galena and sphalerite to Barrow, but the source of the material is unknown (Anderson, 1947, p. 32).

No valuable placer deposits have been found in the northern Alaska region, although, according to Mangus (1953, p. 22), "small amounts of gold have been reported in the drainages of the Firth, Malcolm, and Kongakut Rivers, but are of no commercial value." The Kongakut River rises south of the Romanzof Mountains and flows eastward, then northward to the Arctic Ocean. The Firth and Malcolm are farther east and have most of their courses on the Canadian side of the international boundary.

Four men reported to have been working on the upper Firth River in 1949 and 1950 (Alaska Department of Mines, 1950, p. 51) were on the Canadian side of the international boundary (M. D. Mangus, written commun., Aug. 18, 1970).

NORTHWESTERN ALASKA REGION

The northwestern Alaska region (pl. 1, fig. 18) is the area drained by streams flowing into Kotzebue Sound from the Wulik River on the north to (and including) the Kauk River on the south. It comprises the Kiana, Noatak, Selawik, and Shungnak districts.

The region includes the Baird Mountains, part of the Brooks Range, and diverse parts of lower uplands and lowlands drained mainly by the Noatak, Kobuk, and Selawik Rivers. Most of the

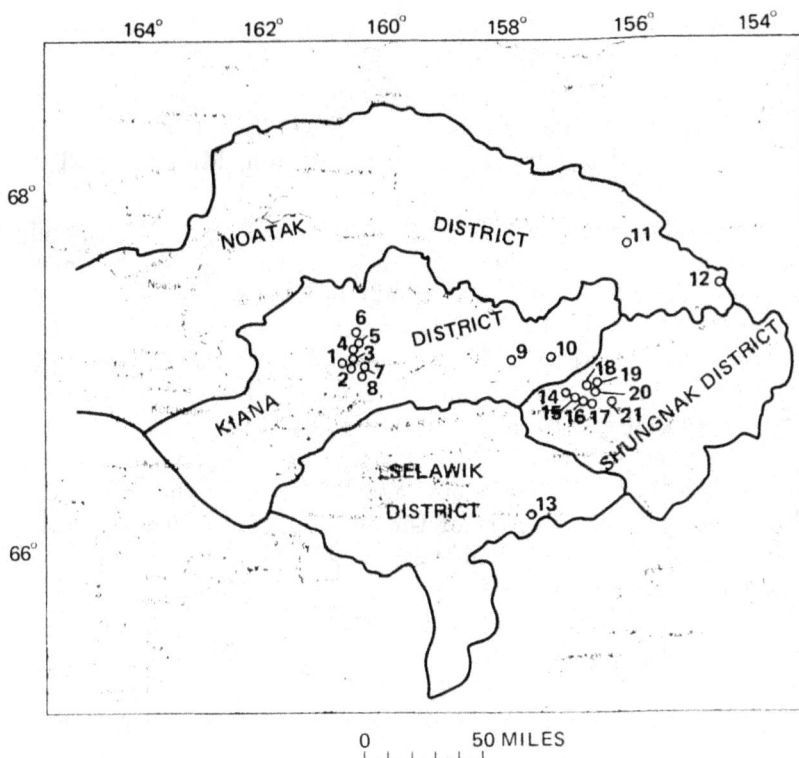

FIGURE 18.—Placer deposits in the northwestern Alaska region.

Kiana district

1. Peluk (Peluck) Creek: Reed (1931b, p. 11).
2. Boldrin (Baldwin) Creek: Reed (1931b, p. 11). Klery Creek: Smith (1913, p. 133–138), Reed (1931b, p. 4–11, 13–14).
3. Joe Gulch: Reed (1931b, p. 12). Klery Creek: Smith (1913, p. 133–138), Reed (1931b, p. 4–11, 13–14).
4. Caribou Creek: Reed (1931b, p. 12, 14). Klery Creek: Smith (1913, 133–138), Reed (1931b, p. 4–11, 13–14).
5. Gold Run Creek: Reed (1931b, p. 12). Klery Creek: Smith (1913, p. 133–138), Reed (1931b, p. 4–11, 13–14).
6. Homestake Creek: Reed (1931b, p. 13).
7. Bear Creek: Reed (1931b, p. 11, 14).
8. Central Creek: Reed (1931b, p. 12–14).
9. Jade Creek: Anderson (1945, p. 5–6, 24–26).
10. Agnes Creek: Smith (1933b, p. 52).

Noatak district

11. Midas Creek: Smith (1913, p. 141–142).
12. Lucky Six Creek: Smith (1913, p. 140–141).

Selawik district

13. Shovel Creek (Purcell Mountain): Alaska Division of Mines and Minerals (1960, p. 70), Miller and Ferrians (1968, p. 11).

Shungnak district

14. Bismark Creek: Reed (1931a, p. 14–15). Shungnak River: Smith (1913, p. 131–132), Reed (1931a, p. 12–16), Anderson (1945, p. 24–26).
15. Cosmos Creek: Anderson (1945, p. 26).
16. Wesley Creek: Anderson (1945, p. 25–26).
17. Dahl Creek: Smith (1913, p. 126–129), Anderson (1945, p. 25–26), Anderson (1947, p. 23), Fritts (1969, p. 26–29).
18. Jay Creek and Ruby Creek: Reed (1931a, p. 18).
19. Pearl Creek: Smith (1934b, p. 53).
20. Riley Creek: Smith (1913, p. 129–130), Reed (1931a, p. 22–24).
21. California Creek: Cathcart (1920, p. 197–198), Anderson (1945, p. 25), Anderson (1947, p. 47), Fritts (1969, p. 28). Lynx Creek: Reed (1931a, p. 27–28), Smith (1942b, p. 64), Fritts (1969, p. 29).

mountains are in rugged eastward-trending groups, 3,000–5,000 feet in altitude, separated by lake-dotted tundra-floored lowlands. A few peaks in the eastern part of the region reach altitudes of more than 8,000 feet.

The following summary of the geology of northwestern Alaska is based on reports and maps by Smith (1913), Patton (1966, 1967), Patton and Miller (1966, 1968), Patton, Miller, and Tailleur (1968), and Fritts (1969).

In the northern part of the region, which is still incompletely mapped, the bedded rocks include limestone and marble, dolomitic limestone (some reefoid), and metamorphosed clastic and volcanic rocks, of early Paleozoic age, that have been complexly folded and displaced by thrust and normal faults. Mafic and ultramafic intrusive rocks of probable Jurassic age cut the older rocks near Misheguk Mountain and north of the Kobuk River. Less disturbed Cretaceous clastic rocks overlie the older units. A Cretaceous granite stock crops out near Kobuk in the Shungnak district. In the southern part of the region, the bedded rocks are mainly Cretaceous clastic and volcanic rocks that were intruded by granitic plutons, some of which are alkalic, and are overlain by Tertiary and Quaternary basalt flows. Most of the lowland areas are covered by thick deposits of Quaternary fluvial, glacial, eolian, and marine sand, silt, and gravel.

Pleistocene ice covered northwestern Alaska except for parts of the Baird Mountains and the southern part of the Selawik district. Only a few cirque glaciers are preserved in the Schwatka Mountains. The region is underlain by continuous permafrost except for an area near Purcell Mountain in the Selawik district and along some streams.

Lodes potentially valuable for copper have been found in the Kiana, Noatak, and Shungnak districts, but only one lode at Bornite (near loc. 18, fig. 18) has been extensively explored (Berg and Cobb, 1967, p. 104–106, fig. 20). Galena, silver, and a little gold are constituents of some lodes. Magnetite-bearing float is common on Iron Mountain near Bornite, but the bedrock source has not been found. Nephrite jade, some of gem quality, occurs with asbestos and may have been mined directly from bedrock in the Jade Mountains near Jade Creek (9, fig. 18), although most of the material produced was stream boulders.

Placer gold was discovered near Shungnak in 1898, on Klery Creek (2–5, fig. 18) in 1909, and, many years later, near Purcell Mountain (13, fig. 18). Calculations based on official records indicate that the total production from the northwestern Alaska region, as of 1960, was about 34,550 fine ounces, but estimates given by Reed (1931b, p. 13–14) and Fritts (1969, p. 28) indicate that the actual total was probably at least double that reported. Jade and quartz crystals are the only mineral commodities other than gold that have been recovered from placer deposits in the region.

KIANA DISTRICT

The Kiana district (fig. 18) includes most of the Baldwin Peninsula and the area drained by the Kobuk River and its tributaries up to (and including) the Ambler River.

Placers on tributaries of the Squirrel River, which flows into the Kobuk at Kiana, have probably accounted for about half of the gold mined in the northwestern Alaska region. Gold was discovered on Klery Creek in 1909 and has been mined from it and some of its tributaries in nearly every year since, generally by one- and two-man operations, although a small dredge was used on the lower part of the main stream (2, 3, fig. 18) for several years. Bedrock is limestone and schist. Much of the gold recovered was fairly coarse and some had fragments of quartz still attached to nuggets, indicating that the gold was derived from quartz veins. Some of the richer deposits may have been formed by reconcentration from an old channel of Klery Creek; magnetite was reported to be the commonest constituent of the concentrates. Gold was found on other tributaries of the Squirrel River, but the amount mined probably was only a few hundred ounces, most of which came from Homestake (6, fig. 18) and Central (8, fig. 18) Creeks.

The only other place in the Kiana district where placer gold was recovered was on Agnes Creek (10, fig. 18), where the little work done was more in the nature of prospecting than mining. Prospect drilling in the valley of the Salmon River, a tributary of the Kobuk about 20 miles east of the Squirrel River (Alaska Department of Mines, 1948, p. 41; Alaska Department of Mines, 1950, p. 45) evidently was not successful, as no mining was reported in later years.

Nephrite boulders from Jade Creek (9, fig. 18) have been collected and shipped for jewelry manufacturing.

NOATAK DISTRICT

The Noatak district (fig. 18) is the area drained by the Noatak River and by coastal streams between its mouth and (but excluding) the Wulik River.

The only recorded mineral production from the district is a few ounces of coarse gold from Lucky Six Creek (12, fig. 18), where bedrock is mainly schist. A nearby lode contains sulfide minerals and a little gold (Berg and Cobb, 1967, p. 105). Small particles of gold found in 1904 on Midas Creek (11, fig. 18) may have come from outwash or from unexplored mountains south of the Noatak River.

SELAWIK DISTRICT

The Selawik district (fig. 18) includes the base of the Baldwin Peninsula and the area drained by streams flowing into Selawik Lake and Eschscholtz Bay between the Kobuk and Kauk Rivers.

The only mineral deposit known in the district is a gold placer on Shovel Creek (13, fig. 18), a small tributary of the Selawik River that drains the northwestern slopes of Purcell Mountain. Two men using nonfloat methods mined there for about 10 years in the 1950's and 1960's. Production is not known and the deposit is now inactive. Shovel Creek crosses the contact between andesitic volcanic rocks of Jurassic(?) and Early Cretaceous age and a quartz monzonite pluton. The gold may have been derived from quartz-tourmaline-sulfide veins near this contact.

SHUNGNAK DISTRICT

The Shungnak district (fig. 18) is the area drained by the Kobuk River and its tributaries above the Ambler River.

All the productive placer mining in the district has been on streams draining the Cosmos Hills, an elongate range about 20 miles long that rises 2,000–3,250 feet above the Kobuk Valley. About half of the gold and most of the jade from the northwestern Alaska region have been recovered from these streams. The hills are underlain by metamorphosed Paleozoic carbonate and clastic rocks, Jurassic mafic volcanic rocks and serpentinite, and, locally, schistose Cretaceous conglomerate, sandstone, and mudstone. A small granite stock intruded Paleozoic rocks in the eastern part of the range. The most likely source of the gold in the placers was quartz veins in the metamorphosed rocks (Patton and others, 1968; Fritts, 1969). Other constituents of placer concentrates are magnetite and chromite, probably at least in part from serpentinite bodies, and native copper and silver, probably from sulfide deposits in Paleozoic carbonate rocks.

Gold was discovered on Dahl Creek (17, fig. 18) in 1898 and soon thereafter on many of the other streams. Most of the production has been from Dahl Creek, which was still being worked in 1968, California and Lynx Creeks (21, fig. 18), and the Shungnak River (14, fig. 18). Other streams listed as sources of gold production include Ruby (18, fig. 18) and Riley (20, fig. 18) Creeks, which drain northward from the Cosmos Hills. Most mining has been done by individuals or small groups.

Nephrite and relatively unsheared serpentinite boulders (Fritts, 1969, p. 29) have been collected from creek gravels and tailings piles and sold for the manufacture of jewelry and objets d'art, for which a heavy demand has developed. Anderson (1947, p. 47) reported that large numbers of quartz crystals were recovered during placer operations on California Creek but did not explicitly state that any were sold. Prospecting was carried on in other parts of the Shungnak district, in particular near Walker Lake (Smith, 1930b, p. 44–45) in the eastern part of the district, but there is no indication that gold was found. In any event, no production was listed from areas not in the Cosmos Hills.

SEWARD PENINSULA REGION

The Seward Peninsula region (pl. 1, fig. 19) includes the Seward Peninsula and the drainage basins of the Buckland, Inglutalik, Ungalik, and Shaktolik Rivers and Egavik Creek. The region comprises the Council, Fairhaven, Kougarok, Koyuk, Nome, Port Clarence, and Serpentine districts.

The Seward Peninsula consists mostly of rounded hills and flat divides 500–2,000 feet in altitude, but there are also isolated groups of rugged glaciated mountains 20–60 miles long and 10 miles wide that rise to peaks 2,500–4,700 feet in altitude. A relatively small part of the region consists of interior lowlands and a narrow coastal plain.

About half of the region is underlain by predominantly Paleozoic schist, gneiss, limestone, and slate, but locally, as in the Kigluaik and Bendeleben Mountains, the rocks may be as old as Precambrian. Most of the region's lode deposits are in the Paleozoic rocks. Mesozoic volcanic and clastic rocks are abundant in the eastern part of the region. The youngest stratified rocks are volcanic ash and basaltic lava flows, some of which may be as old as late Tertiary, but which are mainly Pleistocene in age. One lava flow near Imuruk Lake is not more than a few centuries old.

Most of the pre-Tertiary bedded rocks are in imbricate thrust sheets (Sainsbury, 1969b) that were later folded and displaced by normal faults and were intruded by stocks and batholiths that range in age between 100 and 90 million years as determined by the potassium-argon method on hornblende and biotite (Miller and others, 1966; Sainsbury, 1969a, p. 41). The main outcrop areas of intrusive rocks are in the Kigluaik, Darby, and Bendeleben Mountains; in an area south of Eschscholtz Bay; and near the western tip of the peninsula.

Sand, gravel, and silt thinly mantle bedrock throughout the Seward Peninsula region and in the lowlands form deposits as much as 100 feet thick. The Kigluaik, Bendeleben, Darby, and York Mountains and highland areas east and west of the Kiwalik River in the eastern part of the region were glaciated during the Pleistocene epoch. Ice extended to the Bering Sea from the Kigluaik Mountains at Nome and from the York Mountains in the western part of the peninsula. The peninsula is generally underlain by permafrost.

Metalliferous lodes that were sources of significant amounts of ore, or that contain important resources, include deposits of tin, tungsten, antimony, and beryllium minerals, and gold. Small amounts of bismuth, copper, lead, iron, and other metals have been discovered, but no major deposits of those metals are known.

As of 1961, the Seward Peninsula had been the source of 30.7 percent (about 6,260,000 fine ounces) of the placer gold produced in Alaska and ranked second only to the Fairbanks district of the Yukon River region. Data for the years since 1961 remain confidential. Large-scale mining ceased at the end of the 1962 season. Data are not adequate to apportion the total published production for the region among the districts of the Seward Peninsula, but probably 75–80 percent of the placer gold reported came from the Nome district and most of the rest came from the Kougarok, Council, and Fairhaven districts.

About 2,000 tons of placer tin has been recovered from valleys in the western part of the Seward Peninsula (Berg and others, 1964, p. 119). A little byproduct placer platinum has come from the eastern part of the region and about 1,800 units (31,000 pounds) of WO_3 (tungstic oxide) has come from streams and residual material near scheelite-bearing lodes in the Nome area (Coats, 1944c, pl. 1). An additional unknown, but undoubtedly small amount of scheelite was saved during placer-mining operations for gold in both the eastern and western parts of the peninsula.

COUNCIL DISTRICT

The Council district (fig. 19) is the area drained by the Kwiniuk and Topkok Rivers and intermediate streams flowing into Norton Sound.

Metalliferous lodes scattered through the district have been the sources of minor amounts of silver, lead, gold, and quicksilver (Berg and Cobb, 1967, p. 108–114, fig. 21). Antimony, copper, tin, and zinc minerals have been found, but none has been mined.

Council district

1. Fox River: Smith and Eakin (1911, p. 117). I X L Gulch: Collier and others (1908, p. 237–238).

2. Aggie Creek: Smith (1939b, p. 67–68). Rock Creek: Smith (1929, p. 26).

3. California Creek: Smith (1942b, p. 61–62).

4. Eldorado Creek: Cathcart (1922, p. 206). Herreid (1965a, p. 5–6). Ryan Creek: Collier and others (1908, p. 283, 293), Herreid (1965a, p. 5). Silverbow (Little Anvil) Creek: Collier and others (1908, p. 283, 293).

5. Daniels Creek: Collier and others (1908, p. 283–293), Cathcart (1922, p. 196), Mining World (1941). Koyana Creek: Collier and others (1908, p. 283, 289), Cathcart (1922, p. 185), Herreid (1965a, p. 5). Swede Gulch: Herreid (1965a, p. 5, 8).

6. Golovnin Bay: West (1953, p. 4).

7. Kwiniuk River: West (1953, p. 6).

Fairhaven district

8. Humboldt Creek: Sainsbury and others (1968).

9. Esperanza Creek: Henshaw (1910, p. 366).

10. Chicago Creek: Moffit (1905, p. 67). Kugruk River: Henshaw (1910, p. 369).

11. Alder Creek and Beach: Mendenhall (1902, p. 51), Smith (1930a, p. 34).

12. Dixie Creek: Moffit (1905, p. 64–65).

13. Glacier Creek: Henshaw (1910, p. 369–371).

14. Gold Run (Creek): Henshaw (1910, p. 371), Anderson (1947, p. 45). Trio Creek: Henshaw (1910, p. 371).

15. Duck Creek: West and Matzko (1953, p. 26–27).

16. East and West Clem Creeks: West and Matzko (1953, p. 26–27).

17. Koopuk (Koobuk) Creek (River): Brooks (1925, p. 50).

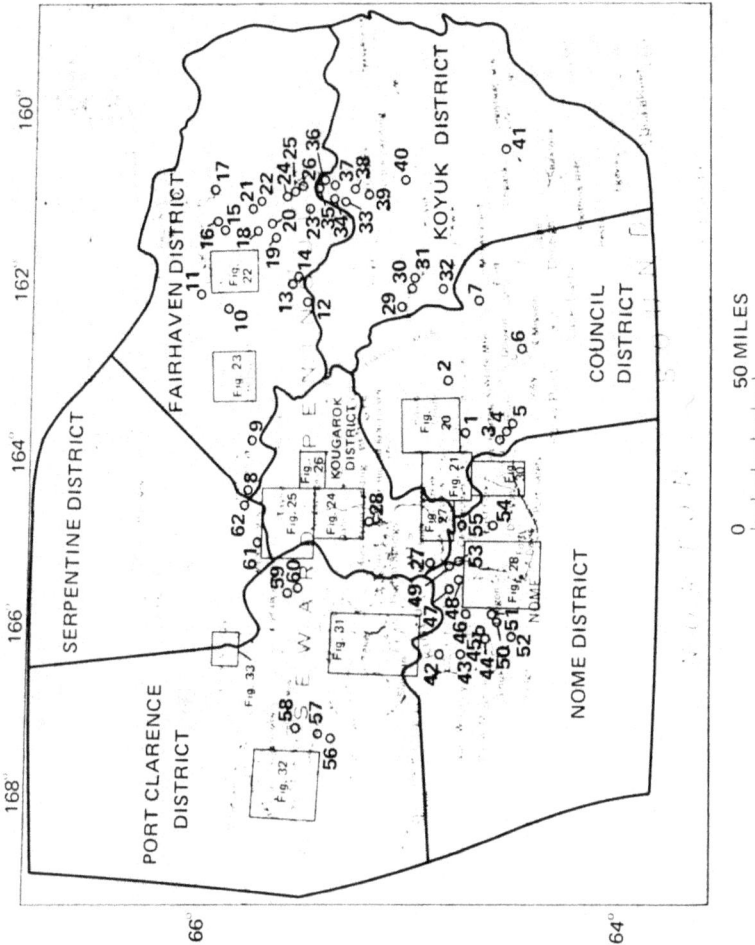

FIGURE 19.—Placer deposits in the Seward Peninsula region.

FIGURE 19.—Continued.

Fairhaven district—Continued

18. Muck Creek: West and Matzko (1953, p. 24–25, 27).
19. Connolly Creek: West and Matzko (1953, p. 25, 27).
20. Spruce Creek: West and Matzko (1953, p. 22, 25, 27).
21. Meinzer Creek: West and Matzko (1953, p. 22, 27).
22. Sugar Loaf Creek: West and Matzko (1953, p. 22, 26–27).
23. Quartz Creek: Smith (1934a, p. 63), Killeen and White (1953).
24. Sheridan Creek: Harrington (1919b, p. 392–394).
25. Bear Creek: Harrington (1919b, p. 392–394). Herreid (1965c, p. 12–14).
26. Cub Creek: Harrington (1919b, p. 392–393), West and Matzko (1953, p. 24–25).

Kougarok district

27. Thompson Creek: Hummel (1961, p. D199).
28. Idaho Creek: Brooks (1901, p. 123).

Koyuk district

29. Otter Creek: Herreid (1965b, p. 5).
30. Grouse Creek: West (1953, p. 3).
31. Camp Creek: Smith and Eakin (1911, p. 115–116).
32. Clear Creek: West (1953, p. 6–7).
33. Bear Gulch: Gault, Black, and Lyons (1953, p. 3–4, 8). Sweepstakes Creek: Harrington (1919b, p. 380–381, 395), Gault, Black, and Lyons (1953, p. 1, 3–9).
34. Spring Creek: Gault, Black, and Lyons (1953, p. 3–4). Sweepstakes Creek: Harrington (1919b, p. 380–381, 395), Gault, Black, and Lyons (1953, p. 1, 3–9).

Koyuk district—Continued

35. Peace River: West and Matzko (1953, p. 24–26).
36. Rock Creek: West and Matzko (1953, p. 22, 24).
37. Anzac Creek: Gault, Black, and Lyons (1953, p. 5–6, 9), West and Matzko (1953, p. 25). Rube Creek: Harrington (1919b, p. 380–381, 394–395).
38. Peace River: Smith and Eakin (1911, p. 114).
39. Eldorado (Little Eldorado) Creek: Harrington (1919b, p. 396, 398). Dime Creek: Harrington (1919b, p. 380–381, 396–398), Anderson (1947, p. 18).
40. Alameda Creek: Smith and Eakin (1911, p. 110–113).
41. Bonanza Creek: Smith and Eakin (1911, p. 105–107), Martin (1920, p. 50). Hopeful Gulch: Anderson (1947, p. 18, 45). Ungalik River: Smith and Eakin (1911, p. 108), Smith (1939b, p. 76).

Nome district

42. Hume Creek: Collier and others (1908, p. 220).
43. Fairview and Tomboy Creeks: Collier and others (1908, p. 218–219).
44. Boulder Creek: Collier and others (1908, p. 216–217).
45. Coal Creek and Washington Creek (Green Gulch): Collier and others (1908, p. 216–217).
46. Rulby Creek: Collier and others (1908, p. 216).
47. Charley Creek: Moffit (1913, p. 133).
48. Stewart River: Smith (1909a, p. 280).
49. Boer Creek: Moffit (1913, p. 76, 100). Divide Creek: Moffit (1913, p. 100).
50. Cripple River: Photointerpretation.

Nome district—Continued

51. Cripple River: Collier and others (1908, p. 210–211). Stella (Slate) Creek: Brooks (1901, p. 96).
52. Quartz Creek: Collier and others (1908, p. 215).
53. Christian Creek: Alaska Department of Mines (1962, p. 57). Dorothy Creek: Moffit (1913, p. 98), Anderson (1947, p. 11).
54. Beaver Creek: Smith (1942b, p. 56–57). Pajara Creek: Alaska Department of Mines (1946, p. 38).
55. Venetia Creek: Collier and others (1908, p. 221–222).

Port Clarence district

56. Lost River: Mulligan (1959b, p. 12–15). Rapid River: Mulligan (1959b, p. 13–14).
57. Cassiterite Creek: Steidtmann and Cathcart (1922, p. 74), Anderson (1947, p. 44), Alaska Department of Mines (1950, p. 53). Lost River: Mulligan (1959b, p. 12–15), Alaska Division of Mines and Minerals (1966, p. 103).
58. York Creek (River) and York Creek, West Fork: Mulligan (1959b, p. 15–17).
59. Goldrun (Gold Run) Creek: Alaska Division of Mines and Minerals (1964, p. 88), Cobb and Sainsbury (1968, p. 5).
60. Budd Creek: Anderson (1947, p. 22), Moxham and West (1953, p. 4, 6), Malone (1962, p. 55). Windy Creek: Eakin (1915c, p. 372), Moxham and West (1953, p. 4).

Serpentine district

61. Dick Creek: Anderson (1947, p. 41, 43–44), Moxham and West (1953, p. 4–6).
62. Hot Springs Creek: Moxham and West (1953, p. 4, 6–11).

Placer deposits in the district may have been discovered as early as 1865–66 on the Niukluk River (2, 14–16, fig. 20) by a party surveying a telegraph route that was planned to connect Europe and America by way of Siberia and the Bering Strait (Collier and others, 1908, p. 13). In 1897 workable placers were discovered on Ophir Greek (8, fig. 20) and the next year a mining district was organized. This, the first gold discovery on the Seward Peninsula, was soon to be followed by major finds near Nome (Jafet Lindeberg, letter to F. L. Hess, quoted in Collier and others, 1908, p. 16–18). During the winter of 1899–1900, a phenomenally rich beach at the mouth of Daniels Creek (5, fig. 19) was discovered and gold worth $600,000 was recovered from it; productive placers were developed on the Casadepaga River (4–10, fig. 21) and a number of its tributaries (Collier and others, 1908, p. 27). By 1900 the three major mining centers of the Council district, the Council, Casadepaga, and Bluff areas, had been discovered and were being intensively developed.

The following discussion of the Council area (fig. 20) is based mainly on descriptions by A. J. Collier and F. L. Hess (Collier and others, 1908, p. 238–257) and Smith and Eakin (1911, p. 117–123).

The Council area is drained by the Niukluk River, a major tributary of the Fish River, and its tributaries below the Casadepaga River. Bedrock is schist and limestone, both of which contain small quartz and calcite veins. Many of the veins contain sulfide minerals and visible gold. While not minable themselves, the veins probably were the source of the gold in the placer deposits.

Workable placers in the present stream gravels and in bench deposits have been mined by hand methods, by hydraulicking, and by dredges. On Ophir Creek, the most important producing stream in the district, the richest deposits were first worked by hand labor and horse-drawn scrapers and then reworked by dredges. In 1968 one of the few gold dredges being operated in Alaska was on Ophir Creek. Magnetite, ilmenite, garnet, pyrite, and hematite, as well as gold, are constituents of the concentrates from most of the creeks. Rutile and scheelite have been reported from Gold-bottom and Warm Creeks (3, fig. 20). Because the smaller creeks often suffered water shortages, those creeks and many bench deposits could be worked only by small-scale methods or after ditches, some several miles long, had been constructed.

In a similar geologic setting south of the Council area, there was sporadic, small-scale mining on Fox River and its tributary I X L Gulch (1, fig. 19).

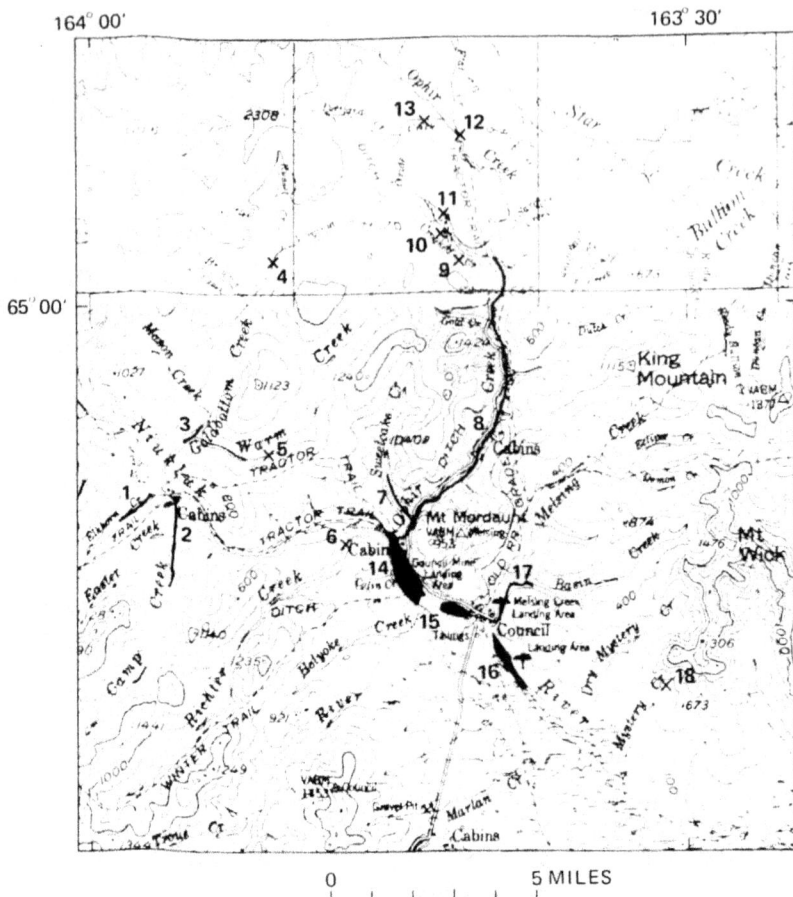

FIGURE 20.—Placer deposits in the Council area.

1. Elkhorn Creek: Collier and others (1908, p. 256-257).

2. Camp Creek: Collier and others (1908, p. 256). Niukluk River: Collier and others (1908, p. 236, 238-239, 263).

3. Goldbottom Creek: Collier and others (1908, p. 254-255), Smith and Eakin (1911, p. 117, 222). Warm Creek: Collier and others (1908, p. 256).

4. Goldbottom Creek: Collier and others (1908, p. 255).

5. Warm Creek: Collier and others (1908, p. 254-256).

6. Richter Creek: Collier and others (1908, p. 263).

7. Sweetcake Creek: Collier and others (1908, p. 250-251).

8. Dutch Creek: Smith and Eakin (1911, p. 120). Ophir Creek: Smith and Eakin (1911, p. 117-121).

9. Balm of Gilead Gulch: Collier and others (1908, p. 254).

10. Albion Creek (Gulch): Collier and others (1908, p. 254), Smith and Eakin (1911, p. 121).

11. Crooked Creek: Moffit (1906b, p. 139), Collier and others (1908, p. 244, 251-253, 262).

12. Ophir Creek: Smith and Eakin (1911, p. 121).

13. Oxide Creek: Collier and others (1908, p. 244).

14-16. Niukluk River: Collier and others (1908, p. 236, 238-239).

17. Basin Creek: Smith and Eakin (1911, p. 118), Smith (1930b, p. 41). Benson Gulch: Smith (1932, p. 46). Melsing Creek: Collier and others (1908, p. 240-242).

18. Mud Creek: Collier and others (1908, p. 240). Mystery Creek: Collier and others (1908, p. 236, 240).

Aggie Creek (a left-limit tributary of Fish River that flows into the main stream 2.9 miles below the stream incorrectly labeled Aggie Creek on most published maps) and its tributary Rock Creek (2, fig. 19), about 13.5 miles east of Council, were the site of mining in the 1930's and early 1940's. The geology of this part of the district has not been studied in the same detail as the area near Council, but bedrock and mineralization probably are similar. All that can be said with certainty is that the deposit on Aggie Creek was large enough to support a dredge from 1938 to 1940 or 1941 and that the dredge tailings are very conspicuous on aerial photographs taken in 1950.

Mendenhall (1901, p. 212) reported fine colors of gold from the bars of Fish River from its mouth to the mouth of Anaconda Creek (now called Pargon River), about 8 miles above Aggie Creek. As later reports did not amplify Mendenhall's original statement, although several referred to it, gold must not be present in minable concentrations.

The following discussion of the Casadepaga area (fig. 21) is based almost entirely on P. S. Smith's detailed descriptions (Smith, 1910) of the regional geology and individual creeks.

The area is drained by the Casadepaga River and American Creek, right-limit headwater tributaries of the Niukluk River. Bedrock is complexly folded and faulted schist, slate, limestone, and greenstone. Quartz veins that contain sulfide minerals and gold are widespread in most bedrock units, especially near schist-limestone contacts (which may be thrust faults) and in slate, but nowhere are they rich enough for a lode mine to have been developed. These mineralized veins, however, were the source of the placer gold in the area. The placer deposits worked in the early 1900's were mainly stream placers on tributaries of the Casadepaga River and bench placers along the main stream (4–10, fig. 21) and its tributaries. Most of the deposits near the mouths of tributaries were reconcentrations of gold from benches along the Casadepaga. On Goose (20–21, fig. 21), Canyon (25, fig. 21), and Lower Willow (31, fig. 21) Creeks both the left-limit bench of the Casadepaga River and benches in the tributary valleys were sites of mining. Heavy minerals in concentrates were mainly magnetite and garnet, which was especially abundant on Ruby Creek (34, fig. 21). Scheelite, cassiterite, and cinnabar, which are fairly common in other placer areas on the Seward Peninsula, have not been reported from the Casadepaga or its tributaries.

American Creek (2, fig. 21) flows into the Niukluk River about 2 miles above the mouth of the Casadepaga River. According to

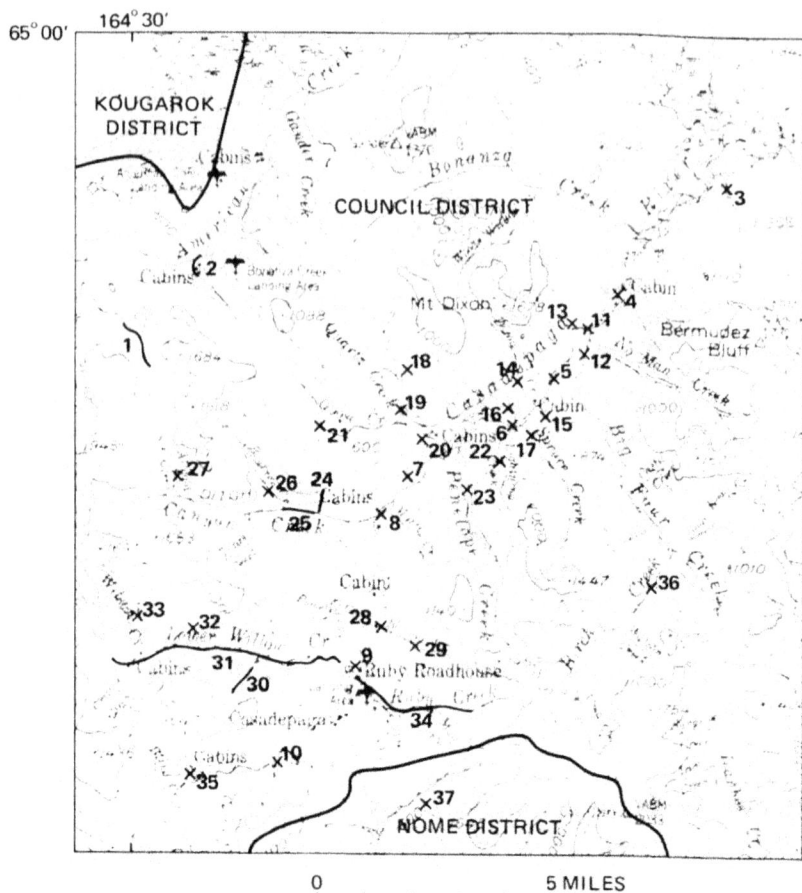

FIGURE 21.—Placer deposits in the Casadepaga area.

1. Auburn Ravine: Smith (1910, p. 204–206).

2. American Creek: Smith (1910, p. 46–47, 204).

3. Puckmummie Creek: Smith (1910, p. 190).

4–10. Casadepaga River: Smith (1910, p. 184–189).

11. No Man Creek: Smith (1910, p. 191).

12. Big Four Creek: Smith (1910, p. 191–192).

13. Dawson Creek: Smith (1910, p. 192).

14. Dixon Creek: Smith (1910, p. 192).

15. Thorp Creek: Smith (1910, p. 192).

16. Dry Creek: Smith (1910, p. 193).

17. Spruce Creek: Smith (1910, p. 193).

18. Gold Moon Gulch: Smith (1910, p. 197).

19. Quartz Creek: Smith (1910, p. 194, 196–197).

20–21. Goose Creek: Smith (1910, p. 194–197).

22. Lightning Creek: Smith (1910, p. 193).

23. Penelope Creek: Smith (1910, p. 193–194).

24. Sunshine Creek: Smith (1910, p. 197, 199).

25. Canyon Creek: Smith (1910, p. 197–199).

26. Boulder Creek: Smith (1910, p. 198).

27. Allgold Creek: Smith (1909b, p. 337).

28–29. Banner Creek: Brooks (1901, p. 107, 109).

30. Ridgeway Creek: Smith (1910, p. 202).

31. Lower Willow Creek: Smith (1910, p. 201–202).

32. Cahill Creek: Smith (1910, p. 202).

33. Wilson Creek: Smith (1909b, p. 336–337).

34. Ruby Creek: Smith (1910, p. 199–200).

35. Moonlight Creek: Mertie (1918b, p. 455).

36. Birch Creek: Collier and others (1908, p. 264).

37. Nugget Gulch: Smith (1910, p. 184).

Smith's (1910, p. 46–47) interpretation, American Creek at one time flowed northwestward into the drainage of the Kruzgamepa River. Stream capture by a tributary of the Niukluk River diverted the headward portion of the American Creek drainage into its present course. The placers in the American Creek basin are all above the postulated point of capture. As in the Casadepaga basin, bedrock is schist and limestone. Mining before 1910 was mainly on Auburn Ravine (1, fig. 21), the major headwater tributary of American Creek, and was severely hampered by water shortages. Concentrates contained garnet, magnetite, ilmenite, and a few grains of cinnabar.

Before about 1909 all mining in the Casadepaga area was by relatively simple methods. Water shortages necessitated the construction of ditches, but not on as large a scale as in many other parts of the Seward Peninsula. In 1909 a dredge was built on Goose Creek (20, fig. 21) in a place where so much water was lost in cavernous limestone that there was not enough in the stream to float the boat. Although this dredge probably never produced anything, dredges on other streams, including the Casadepaga River, Lower Willow Creek, Ruby Creek, and American Creek near the mouth of Game Creek, were successfully operated before World War II. Since the war there has been practically no mining in the area.

Near Bluff, an area underlain by limestone and schist, Daniels Creek (5, fig. 19) has been the principal producing stream. Its alluvium and two generations of beach placers near its mouth yielded most of the 70,000–75,000 ounces of placer gold recovered in the area by 1920. The stream placer in Daniels Creek is unique in that it formed in a collapsed cavern in limestone (D. M. Hopkins, oral commun., April 2, 1970). Gold lodes in schist near Daniels and Koyana Creeks (5, fig. 19) and a cinnabar lode near Swede Gulch (5, fig. 19) (Cathcart, 1922, p. 186–197; Anderson, 1947, p. 33; Herreid, 1965a) were the sources of the gold and cinnabar in the placers of the area. Scheelite also has been reported from Daniels Creek.

Shortage of water hampered mining until a ditch brought water about 20 miles from the headwaters of the Klokerblok (Kocheblok) River to Daniels Creek. From 1903, when the ditch became operative, until at least as recently as 1956 (Alaska Dept. Mines, 1956, p. 84), there was mining on Daniels Creek or on the beach at its mouth. Of particular interest was a scraper arrangement that permitted winter mining of a submarine channel off the mouth of the creek through a slot cut in the sea ice. The most recent mining was on Eldorado Creek (4, fig. 19), where a small dredge was in operation when Herreid (1965a, p. 6) visited the Bluff area in 1964. Other streams near Bluff where placer gold has been mined are Little Anvil (Silverbow) (4, fig. 19) and California (3, fig. 19) Creeks.

During the 1960's private companies prospected extensively for submarine gold placers in the Bering Sea near Bluff; they have not made public their findings.

The only other places in the district where possibly valuable minerals have been reported from surficial deposits are near Kwiniuk River (7, fig. 19) and Golovnin Bay (6, fig. 19). In the Kwiniuk River basin the Geological Survey found allanite, scheelite, chalcopyrite, and rutile in a sample collected from a small tributary stream that drains a contact zone between granite and Paleozoic sedimentary rocks.There is an unconfirmed report that cinnabar was found on the Kwiniuk River (Anderson, 1947, p. 34). A sample from near Golovnin Bay contained topaz, allanite, scheelite, and an unidentified uranium-titanium-niobium mineral The material came from slopewash on granite near its contact with an older metamorphosed granitic rock.

FAIRHAVEN DISTRICT

The Fairhaven district (fig. 19) is the area drained by the Buckland and Goodhope Rivers and intermediate streams entering Kotzebue Sound and Eschscholtz Bay.

A few lode deposits in the district have been investigated, mainly as possible sources of silver and lead (Berg and Cobb, 1967, p. 114–115, fig. 22). Only one was productive.

Placer gold was discovered in the Inmachuk River area (fig. 23) in 1900, but little mining was done there for several years, as very rich finds made in 1901 on Candle Creek (3, fig. 22) drew practically all of the men in the northeastern Seward Peninsula. Bear Creek (25, fig. 19) was staked in 1901.

Bedrock in the Candle area (fig. 22) is mainly poorly exposed Paleozoic schist and limestone covered by Tertiary(?) and Quaternary basaltic lava flows east of the Kiwalik River (Patton, 1967). At least one granitic dike cuts the Paleozoic rocks on Candle Creek (Henshaw, 1909, p. 365). The source of the gold in the placers was gold-bearing quartz veins in the old rocks, particularly the schist (Moffit, 1905, p. 65–66). Moffit found pyrite and rutile in concentrates, and later studies (Gault, 1953) determined that magnetite, ilmenite, various sulfide minerals, garnet, and zircon are present. A radioactive mineral tentatively identified as uraninite-thorianite (uranothorianite?) was found in a single sample collected in 1917, but no significant radioactivity was

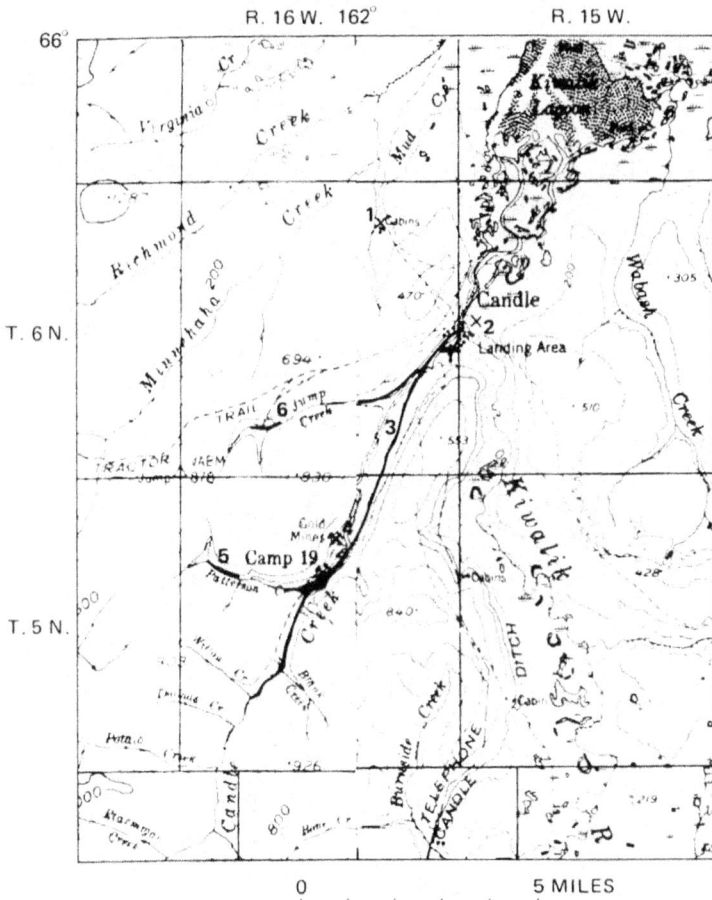

FIGURE 22.—Placer deposits in the Candle area.

1. Minnehaha Creek: Mendenhall (1902, p. 51).
   Mud Creek: Anderson (1947, p. 31, 34).
2. Kiwalik River: Gault (1953, p. 11).
3. Candle Creek: Henshaw (1909, p. 364-368),
   Harrington (1919b, p. 391-392), Anderson (1947, p. 31), Gault (1953). Jump

Creek: Gault (1953, p. 14). Patterson
Creek: Moffit (1905, p. 61).

4. Candle Creek: Gault (1953).
5. Patterson Creek: Smith (1942b, p. 59).
6. Jump Creek: Henshaw (1909, p. 364).

found in any other concentrate from the area. Cinnabar and galena have been reported from Mud Creek (1, fig. 22).

Most of Candle Creek (3, 4, fig. 22) and its tributaries, Jump (3, 6, fig. 22) and Patterson (3, 5, fig. 22) Creeks, were mined by simple methods until many ditches had been constructed. The longest ditch tapped a tributary of the Kiwalik River about 25 airline miles from Candle. Most of the creek gravels along Candle Creek and in an area in the Kiwalik River flats below Candle (2, fig. 22) were dredged. Large hydraulic plants also worked bench gravels along Candle Creek. By 1967 mining in the area had dwindled to a single two-man operation on Mud Creek.

For a few years, beginning in 1908 or 1909, there was considerable placer mining on Glacier (13, fig. 19) and Gold Run (14, fig. 19) Creeks, tributaries of the Kiwalik River that drain a highland area underlain by schist, limestone, and granite. Kyanite, scheelite, and wolframite made up a large part of a concentrate sample from Gold Run Creek. Activity in this area, reported annually from 1927 until World War II and as recently as 1955, probably was more in the nature of prospecting than mining.

The following discussion of the Inmachuk River area (fig. 23) is based largely on Moffit's report (1905) on the Fairhaven placers and Hopkins' description (1963) of the neighboring Imuruk Lake area. Paleozoic schist and crystalline limestone are the most common rock types in the map area (fig. 23), although there are Mesozoic granitic plutons not far to the south. Gravel and basaltic lava flows, some not more than a few hundred years old, cover much of the area. Lava flows form a rimrock overlooking parts of the valleys of the Inmachuk (6, fig. 23) and Pinnell (4, fig. 23) Rivers and Hannum Creek (2, fig. 23). The gold in the placers of these streams and their tributaries probably was derived from the quartz-calcite veins that are abundant in schist bedrock. Cassiterite has been found in American (5, fig. 23), Old Glory (4, fig. 23), and Hannum Creeks, and lead minerals have been found in the Hannum Creek drainage, where there are galena-bearing lode deposits (Moffit, 1905, p. 54; Mulligan, 1965b). Dredge concentrates from the Inmachuk River near the point of the farthest downstream mining contained cinnabar pebbles as large as half an inch in greatest dimension.

Large-scale operations in the area did not become possible until 1908, when the Fairhaven ditch, which brought water more than 40 miles from Imuruk Lake, was completed. Most of the gold mined between 1912 and 1963 came from dredges on the Inmachuk and Pinnell Rivers. Since then activity has been on a smaller scale; some of the production has come from an old channel of the Inmachuk River that had been buried by later lava flows. Similar gravels beneath lava flows high on a valley wall of Perry Creek (7, fig. 23) did not yield much gold.

The major stream between the Kiwalik and Inmachuk drainages is the Kugruk River. In 1903 a little placer mining was reported on Dixie Creek (12, fig. 19), a small stream flowing into Independence Creek, one of the principal headwater tributaries of the Kugruk River. Farther downstream on the Kugruk River, auriferous gravel was found on Chicago Creek and nearby on the

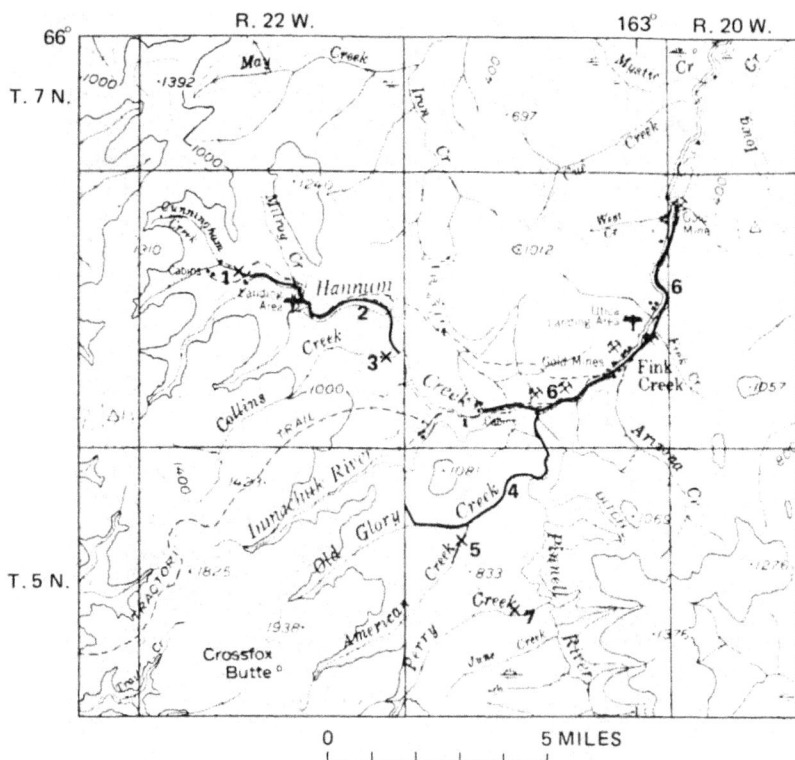

FIGURE 23.—Placer deposits in the Inmachuk River area.

1. Cunningham Creek: Moffit (1905, p. 53–54), Anderson (1947, p. 31).
2. Hannum Creek: Moffit (1905, p. 51–54), Anderson (1947, p. 31, 41). Milroy Creek: Anderson (1947, p. 31).
3. Collins Creek: Moffit (1905, p. 54).
4. Nelson Gulch: Moffit (1905, p. 56–57). Old Glory Creek: Moffit (1905, p. 54–56). Pinnell River: Moffit (1905, p. 57–58), Hen-

shaw (1910, p. 368).
5. American Creek: Moffit (1905, p. 57), Anderson (1947, p. 41).
6. Discovery Gulch: Smith (1941b, p. 63). Inmachuk River: Moffit (1905, p. 58–60), Anderson (1947, p. 34), Hopkins (1963, p. C32).
7. Perry Creek: Moffit (1905, p. 58), Hopkins (1963, p. C32, C94).

Kugruk itself (10, fig. 19). A dredge was operated for a few years on the Kugruk River before being moved to another area in 1916. Mining, including the installation of another dredge that lies abandoned in the river (D. M. Hopkins, oral commun., April 2, 1970), was reported in several later years, but the amount of gold recovered probably did not make up a significant part of the production of the Fairhaven district.

The only other place between the Kiwalik and Goodhope Rivers where placer mining has been reported is the beach at the mouth of Alder Creek (11, fig. 19), where a few hundred ounces of gold was recovered in 1902 and 1903. The gold, probably of local derivation, lay on schist bedrock beneath about 1 foot of beach gravel. A little gold is reported to have been mined from Alder Creek in 1927.

Two streams in the Goodhope River basin have been sites of placer activity. Gold was discovered on Esperanza Creek (9, fig. 19) in 1908, and there may have been some mining in 1909. Fairly large-scale mining for gold was carried on in the headwaters of Humboldt Creek (8, fig. 19) for many years, but operations were hampered by the large amounts of coarse cassiterite that clogged sluice riffles. Humboldt Creek rises near a granite pluton similar to bodies in western Seward Peninsula with which lode and placer tin are associated. According to Sainsbury and others (1968), the remaining creek gravels might be dredged at a profit if both gold and cassiterite were recovered.

The divide between the Buckland and Kiwalik Rivers is underlain by Jurassic(?) and Lower Cretaceous volcanic rocks and by granitic plutons of middle Cretaceous age. Tertiary(?) and Quaternary basaltic lava flows were extruded over a surface of moderate relief that had been developed on the older rocks (Patton, 1967).

In 1947, West and Matzko (1953) visited most of the streams in the Buckland-Kiwalik area and found uranothorianite, though not in commercially recoverable amounts, and traces of other radioactive minerals in concentrates from the gravels of many streams draining areas underlain by granitic rocks near Clem Mountain (15, 16, fig. 19) and Sugar Top (18–22, fig. 19). They identified scheelite in a sample from Muck Creek (18, fig. 19) and found evidence that nearby creeks had been prospected or mined for placer gold.

In the early 1920's, a little gold was reported to have been mined from Koopuk (Koobuk) Creek (17, fig. 19), a small tributary of the Buckland River that drains an area underlain entirely by Cenozoic lava flows (Patton, 1967). The source of the gold is unknown.

Farther south along the Buckland-Kiwalik divide, placer gold has been mined in the valleys of Bear Creek (24–26, fig. 19), a tributary of the Buckland River, and Quartz Creek (23, fig. 19), which flows into the Kiwalik. Bedrock includes Mesozoic volcanic rocks and dikes that are probably related to a pluton exposed nearby at Granite Mountain. Bear Creek was staked in 1901; it has been sporadically mined and prospected since that time. Concentrates contained gold, magnetite, ilmenite, hematite, chrome spinel, garnet, and platinum, some of which was recovered and marketed. Uranothorianite was identified in samples from Cub Creek (26, fig. 19), a tributary from which gold has been mined.

The Quartz Creek drainage basin is geologically similar to that of Bear Creek except that some headwater tributaries cross the contact zone around the Granite Mountain pluton. Gold and platinum have been recovered from its gravels, and uranothorianite and other radioactive minerals have been identified in the heavy-mineral fractions of concentrate samples.

KOUGAROK DISTRICT

The Kougarok district (fig. 19) includes the area drained by the Kaviruk, Kuzitrin, and Kruzgamepa (Pilgrim) Rivers, all of which drain into Imuruk Basin.

The district contains lodes from which lead, silver, and copper ore were shipped, as well as deposits explored for gold, tungsten, lead, silver, and copper (Berg and Cobb, 1967, p. 115, 118–119, fig. 21).

In 1900, placer gold was discovered in the Kougarok Valley (figs. 24 and 25), in small streams tributary to the Noxapaga River (fig. 26), and in the Iron Creek area (fig. 27) (Collier and others, 1908, p. 306, 314; Smith, 1909b, p. 303). Despite two rushes from Nome annual production remained small until 1905, when about 19,350 ounces of gold was reported from the district (Collier and others, 1908, p. 37). Because the fieldwork on which the most complete geologic reports on the Kougarok district (Collier and others, 1908; Smith, 1909b) are based was done before development had progressed greatly, the data, including the location of some of the placer mining operations are incomplete. Moreover; geographic names were so haphazardly applied and changed that existing records are confusing. In determining the sites of mining on the Kougarok River and some other streams, I relied on the interpretation of aerial photographs taken in 1950 and 1955 as well as on published reports. As a result, I may have missed some areas and may have overemphasized others.

The Kougarok district is underlain by probably Precambrian (C. L. Sainsbury, oral commun., March 13, 1970) slate, schist, and schistose limestone, small bodies of altered mafic intrusive rock, and granite stocks. The area is marked by major thrust faults, but the main placer deposits are close to, and downstream from, gold-bearing, hydrothermally altered, north-trending normal faults that postdate the thrusts (Sainsbury and others, 1968, p. 2, 4).

Hopkins (1963, p. C29–C34, C42–C43) described an old gravel as well as much younger stream and low-terrace gravels associated with the present drainage system. Both units are auriferous, but the richest deposits are in the younger gravels, and most of the production has been from them.

Large-scale mining was not possible in the Kougarok Valley (figs. 24 and 25) until many ditch systems had been constructed, as the flow of most of the streams was either too small or too irregular for any but the simplest mining methods. None of the ditches rivalled in length those dug as parts of major integrated operations in other districts of the Seward Peninsula, but several were 3 to 10 miles long and permitted large-scale hydraulick-

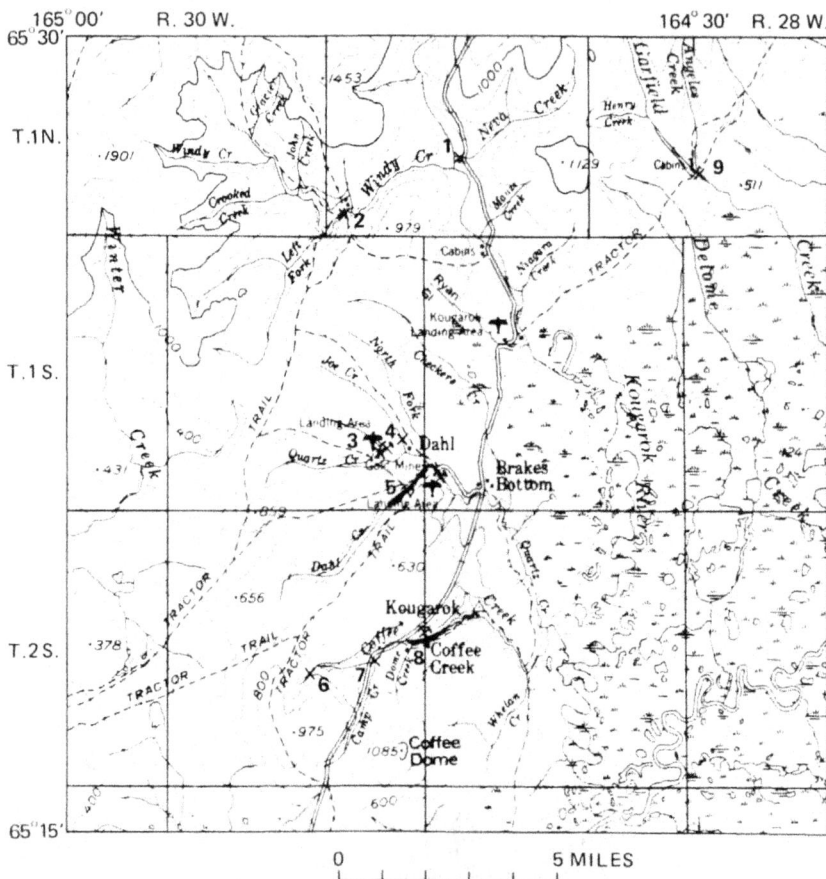

FIGURE 24.—Placer deposits in the southern Kougarok River area.

1. Neva Creek: Collier and others (1908, p. 321).
2. Anderson Gulch: Collier and others (1908, p. 320–321). Windy Creek: Collier and others (1908, p. 320–321).
3. Atlas Creek: Photointerpretation.
4. Joe Creek: Collier and others (1908, p. 311–312).
5. Dahl Creek: Collier and others (1908, p. 310–312). Quartz Creek: Collier and others (1908, p. 306, 311–312).
6–7. Coffee Creek: Collier and others (1908, p. 313).
8. Coffee Creek: Collier and others (1908, p. 313), Anderson (1947, p. 34). Eagle Gulch: Smith (1933b, p. 40). Wonder Gulch: Smith (1934a, p. 45), Anderson (1947, p. 28, 34).
9. Garfield Creek: Collier and others (1908, p. 313–314).

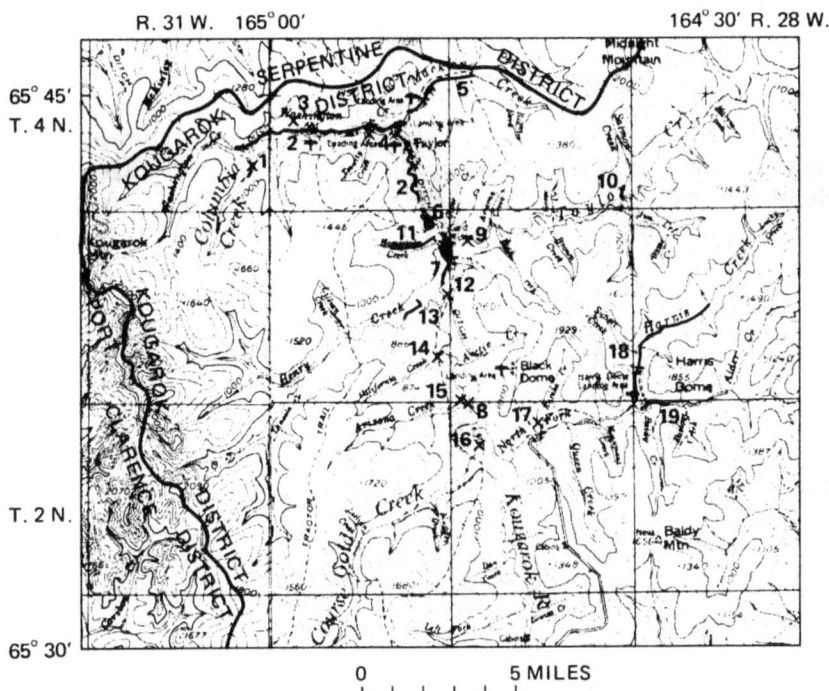

FIGURE 25.—Placer deposits in the northern Kougarok River area.

1. Columbia Creek (location approx.): Collier and others (1908, p. 326).
2. Kougarok River (including Washington Creek): Collier and others (1968, p. 315–320) and photointerpretation.
3. Mascot Gulch (location approx.): Collier and others (1908, p. 320).
4. Trinity Creek: Brooks (1907b, p. 179).
5. Macklin Creek: Smith (1942b, p. 58) and photointerpretation.
6–8. Kougarok River: Collier and others (1908, p. 306–309, 315–320) and photointerpretation.
9. Taylor Creek: Collier and others (1908, p. 324–325).
10. Solomon (Salmon) Creek: Collier and others (1908, p. 308, 325).
11. Homestake Creek: Collier and others

(1908, p. 325–326), Anderson (1947, p. 43).
12. Dreamy Gulch: Collier and others (1908, p. 324).
13. Henry Creek: Collier and others (1908, p. 324).
14. California Creek: Collier and others (1908, p. 324).
15. Arizona Creek: Collier and others (1908, p. 324), Smith (1909a, p. 296).
16. Coarse Gold Creek (location approx.): Collier and others (1908, p. 323).
17. Eureka Creek: Collier and others (1908, p. 321).
18. Harris Creek: Collier and others (1908, p. 321–323), Anderson (1947, p. 31).
19. North Fork: Collier and others (1908, p. 321, 323).

ing. On the Kougarok River (including Washington Creek) (2, 6–8, fig. 25) dredges accounted for most of the reported production. Elsewhere, particularly on Dahl (5, fig. 24), Harris (18, fig. 25), and Macklin (5, fig. 25) Creeks, as well as in bench deposits along the Kougarok River, there were large nonfloat operations. Most of the gravel mined was in stream or bench deposits of the present drainage system, although gold values were found in a body of old gravel in the Quartz Creek basin (5, fig. 24) and in small residual placers in Mascot Gulch (3, fig. 25) and near the head of Coffee Creek (6, fig. 24).

Few data on the mineral composition of cencentrates from the Kougarok Valley have been published. Pyrite, magnetite, and hematite were reported from deposits on the Kougarok River and North Fork (19, fig. 25), cassiterite from Mascot Gulch, scheelite from Homestake Creek (11, fig. 25) and tributaries of Quartz Creek (5, fig. 24), and cinnabar from Coffee Creek and Wonder Gulch (8, fig. 24). Lead minerals were identified in concentrates from Wonder Gulch.

Colors of gold were found in Idaho Creek (28, fig. 19), a small stream that enters the Kuzitrin River a few miles west of the Kougarok River, but no mining has been reported.

The Noxapaga area (fig. 26) is geologically similar to the Kougarok Valley. Gold was discovered on Goose Creek (6, fig. 26) in 1900, and production was reported intermittently for many years from Goose Creek and neighboring small tributaries of the Noxapaga River. The most recent mining reported was on Grouse Creek (3, fig. 26) in 1957. A small patch of old gravel on the Noxapaga River (7, fig. 26) is sparsely auriferous, but there is no record that it was ever mined commercially. Data on produc-

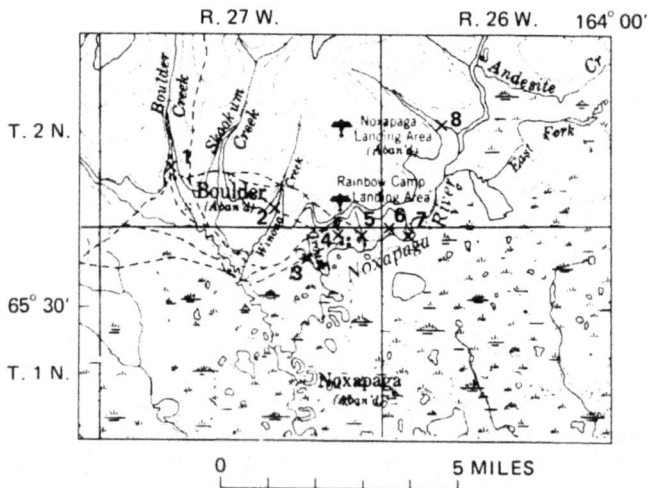

FIGURE 26.—Placer deposits in the Noxapaga River area.

1. Boulder Creek: Collier and others (1908, p. 341).
2. Winona Creek: Hopkins (1963, p. C94).
3. Grouse Creek: Hopkins (1963, p. C94).
4. Black Gulch (Creek): Hopkins (1963, p. C42, C92).
5. Buzzard Gulch (Creek): Hopkins (1963, p. C94).
6. Goose Creek: Collier and others (1908, p. 314), Hopkins (1963, p. C94).
7. Noxapaga River: Hopkins (1963, p. C94).
8. Frost Creek: Hopkins (1963, p. C94).

tion from streams in the Noxapaga area are fragmentary, but the area probably accounted for only a small fraction of the district total.

Garfield Creek (9, fig. 24) is a tributary of the Kuzitrin River between the Kougarok and Noxapaga Rivers. Gold placers in the upper part of its course yielded about $25,000 in coarse gold from a shallow pay streak in the creek bed in 1900–1901. Since then, a little mining was reported for 3 years, but no work is known to have been done since World War I.

The Iron Creek area (fig. 27), which comprises the basins of Iron Creek and a few smaller north-flowing tributaries of the Kruzgamepa (Pilgrim) River, was the source of most of the gold

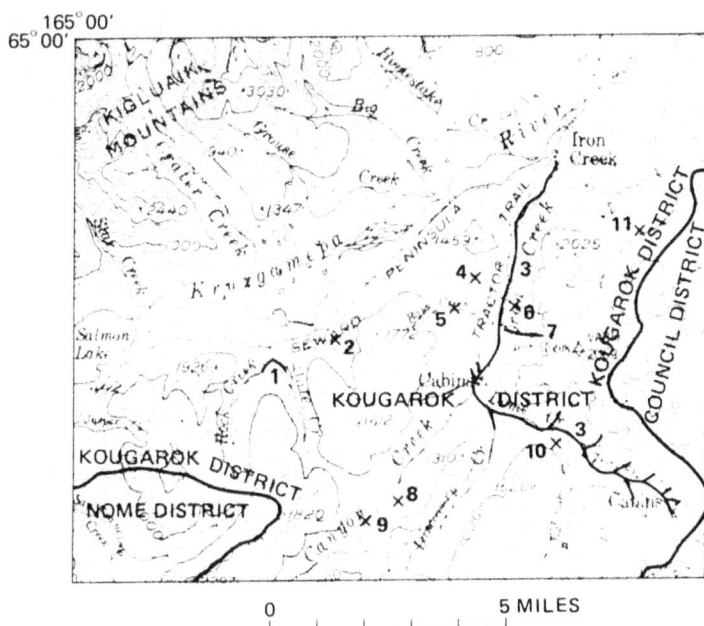

FIGURE 27.—Placer deposits in the Iron Creek area.

1. Rock and Slate Creeks: Smith (1909b, p. 320–321).
2. Willow Creek: Smith (1909b, p. 321).
3. Discovery Creek: Smith (1909b, p. 330). Adventuress, Dividend, and Hardluck Creeks, Hobo Gulch, Left Fork, Oversight and Penny Creeks, Ready Bullion, and Shoal Creeks: Smith (1909b, p. 331). Iron (Dome, Telegram, Telegraph) Creek: Smith (1909b, p. 322–327), Smith (1941b, p. 61) Rabbit Creek: Smith (1909b, p. 329). Rocky Creek: Smith (1907, p. 163).
4. Barney Creek: Smith (1909b, p. 327).
5. Bobs Creek: Smith (1909b, p. 327–328).
6. Easy (Eagle) Creek: Smith (1909b, p. 328, 340–341).
7. Benson Creek: Smith (1909b, p. 328–329), Cathcart (1922, p. 210).
8. El Patron Creek: Smith (1909b, p. 329–330).
9. Canyon Creek: Smith (1909b, p. 329–330).
10. Chickamin Gulch: Smith (1909b, p. 341–342).
11. Sherrette Creek: Smith (1909b, p. 331–333).

produced in the district exclusive of the Kougarok Valley. Workable placers were discovered in 1900 and have been mined by various methods in nearly every year since. The area is geologically similar to the Kougarok Valley. Sulfide-bearing lodes have been explored as possible sources of lead, silver, copper, and gold (Cathcart, 1922, p. 208–217; Asher, 1969). Placer deposits are in both the present stream gravels and older bench gravels, some of which contributed to the richer stream placers where tributaries had cut through them.

Iron Creek (3, fig. 27) was given different names in different parts of its course. This, coupled with an uncommonly thorough job of naming tributary streams and gullies, has resulted in considerable confusion in records. Most of the gold recovered came from the main stream in the parts called Iron Creek and Dome Creek and from Benson Creek (7, fig. 27), a tributary that heads in a ridge where there are several lode deposits. Most mining was by simple methods, as water is scarce; in many places where a stream crosses carbonate rocks, it flows underground. A dredge was installed on Iron Creek in 1939 and operated for at least 2 years.

Slate Creek, whose principal tributary is Rock Creek (1, fig. 27), flows into the Kruzgamepa River about 7 miles above the mouth of Iron Creek. These creeks and neighboring Willow Creek (2, fig. 27) were sites of placer mining before 1915. Some of the gold in Slate Creek was derived from an altered dike, crushed samples of which contained enough gold to be recovered by panning (Chapin, 1914c, p. 405).

Heavy minerals in placer concentrates from the Iron Creek area generally include magnetite, ilmenite, and garnet. Cinnabar was reported from the Dome Creek segment of Iron Creek; iron and copper sulfide minerals were reported from Sherrette Creek (11, fig. 27).

Near Thompson Creek (27, fig. 19), Hummel (1961) found a wide belt of calc-silicate rock along the contact between a thick gneissic granite sill of Paleozoic or Mesozoic age and interbedded schist and marble. The heavy-mineral fraction of a sample collected downstream from this contact contained scheelite.

KOYUK DISTRICT

The Koyuk district (fig. 19) is the part of the Seward Peninsula region drained by streams flowing into Norton Bay and Norton Sound between and including Egavik Creek and the Tubutulik River.

Copper, gold, silver, and antimony lodes are known in the Koyuk district, but none has been productive (Berg and Cobb, 1967, p. 119, fig. 22).

Gold was discovered and claims were staked on Bonanza Creek (41, fig. 19) in 1899 (Smith and Eakin, 1911, p. 105) and on Sweepstakes (33–34, fig. 19) and neighboring creeks about 10 years later (Harrington, 1919b, p. 380). Gold in paying quantities was found on Eldorado and Dime Creeks (39, fig. 19) in 1915. Production from the Koyuk district remained fairly steady, though not large, until after World War II. By 1968, however, mining had dwindled to a three-man nonfloat operation on the Ungalik River (41, fig. 19).

The area near Bonanza Creek and the Ungalik River is underlain by Cretaceous sedimentary rocks cut by sulfide-bearing granitic dikes that may be related to a pluton exposed at Christmas Mountain about 6 miles to the east. Mining of creek and bench placers on Bonanza Creek was hampered by a severe shortage of water that could be alleviated only by pumping from the Ungalik River, but the deposits were rich enough to be worked, even by such expensive methods. The gold is of local origin, as nuggets with vein quartz still attached were common in sluiceboxes. Heavy minerals other than gold in the concentrates included magnetite, ilmenite, scheelite, and stibnite. Scheelite was so plentiful that in 1918 the production of a few pounds of concentrate was reported. Bismuthinite, scheelite, and wolframite were found in neighboring Hopeful Gulch (41, fig. 19). Dredges installed on the Ungalik River near the mouth of Bonanza Creek and a short distance farther downstream in 1938 accounted for much of the production from the district for a few years.

Colors of gold are rumored to have been found in creeks that head near Christmas Mountain and in tributaries of the Inglutalik River, a stream that flows into the head of Norton Bay, but no mining was ever reported from any of them.

Dime Creek and its headwater tributary Eldorado Creek flow near a probably faulted contact between Paleozoic recrystallized carbonate rocks and Jurassic(?) and Cretaceous andesitic volcanic rocks that were intruded by small mafic and ultramafic plutons (Harrington, 1919b, p. 373; Patton, 1967). In 1915, the first prospectors in the area staked claims on Eldorado Creek, leaving what proved to be richer and more extensive creek and bench placers along Dime Creek for latecomers. Mining was reported in practically every year from 1915 until as recently as 1952. A dredge operated for a few years on Dime Creek, but much of the mining was by drifting and other nonmechanized methods. In addition to gold, platinum, derived either from the andesitic volcanics or from mafic or ultramafic bodies intruded into them, has been recovered from the placers of Dime Creek. Other heavy minerals reported from concentrates include magnetite, chromite, a little rutile, and a few garnets.

The only other area in the Koyuk district where there has been significant placer mining is Sweepstakes Creek a tributary of the Peace River that drains the southwestern part of the Granite Mountain pluton. Geologically, the area is similar to that in the drainage basins of Bear and Quartz Creeks in the Fairhaven district. Sweepstakes Creek was staked in 1909, and in most years prior to 1965, gold worth a few thousand dollars was produced from low-bench deposits. Platinum has been recovered from Bear Gulch (33, fig. 19) a small tributary of Sweepstakes Creek from the west, and from Sweepstakes Creek below Bear Gulch. Uranothorianite, hydrothorite, magnetite, ilmenite, hematite, chrome spinel, and garnet were also identified in concentrates. Samples collected by West and Matzko (1953) in an area near the head of a small headwater tributary of the Peace River (35, fig. 19) contained bismuth, iron and copper sulfides, chromite, various uranium and thorium minerals, gold, galena, sphalerite, molybdenite, scheelite, magnetite, and ilmenite. No lode sources were found in the area sampled, which covers the contact zone between the Granite Mountain pluton and volcanic rocks. Gold has been found in other small tributaries of both Sweepstakes Creek and the Peace River, but actual production was reported only from Rube Creek (37, fig. 19). Concentrate samples from some of the creeks were slightly radioactive.

Mendenhall (1901, p. 212–213) and Smith and Eakin (1911, p. 110–116) reported prospecting and possibly a little mining on the Tubutulik River and other creeks that drain an area in the southwestern part of the Koyuk district where bedrock is schist and limestone. Of these occurrences, only those on Alameda (40, fig. 19) and Camp (31, fig. 19) Creeks could be located closely enough to plot on the placer map (fig. 19). Gold was too scarce to support mining operations in any of these creeks.

West (1953) reported signs of old mining on Grouse Creek (30, fig. 19), which follows the contact between a granite pluton and Paleozoic carbonate and metamorphic rocks. He also collected samples in the vicinity of Clear Creek (32, fig. 19), in which he found allanite, an unidentified niobium-titanium-calcium-uranium mineral, ilmenite, hematite, and traces of cassiterite and scheelite.

Bulldozer cuts in alluvium at Otter Creek (29, fig. 19) exposed material in which there was a little cassiterite. Bedrock in this area is mainly Paleozoic(?) schist and marble intruded by small granite and diorite bodies.

NOME DISTRICT

The Nome district (fig. 19) is the part of the Seward Peninsula drained by the Solomon River and other streams flowing into Norton Sound and the Bering Sea as far west as Cape Douglas.

Gold and antimony have been produced from lode deposits in the district and tungsten concentrates have been produced from residual material above scheelite-bearing lodes near Nome. Lodes possibly valuable for other metals, including iron, copper, bismuth, molybdenum, lead, and zinc, are known in the Nome district, but none has been worked successfully (Berg and Cobb, 1967, p. 119–128, fig. 21).

Placer gold was discovered in 1898 on bars of the Snake River (60, 61, fig. 28) and in the gravels of Anvil Creek (30, fig. 28) and some of its tributaries. The following year saw the beginning of a stampede exceeded in North America only by the California and Klondike gold rushes. Claim jumping on the creeks and general lawlessness, compounded by an absence of civil authority, created a stiuation finally relieved by the discovery of gold in the beach gravels near Nome. As claims could not be staked on the beach, it was a daily case of first come, first served, and there was enough beach to accommodate nearly everyone. A. H. Brooks (in Collier and others, 1908, p. 13–39) recounted in detail the early history of discovery and mining on the Seward Peninsula, and Moffit (1913, p. 65–68) summarized the history of the Nome area. The other part of the district where there was extensive mining was the Solomon River valley (fig. 30), where both placer and lode-gold deposits were being worked by 1900.

Bedrock in the Nome district is mainly Paleozoic and Precambrian(?) schist, displaced by at least two generations of folds and faults and intruded by now-altered mafic rock bodies and small granitic plutons. The lodes, most of them spatially and probably genetically related to the younger folds and faults, are mainly quartz veins and veinlets, many of which contain feldspar or calcite, that carry various combinations of free gold, metallic sulfides, and scheelite. The placer deposits in turn were derived from the lodes. Further information on the regional bedrock geology and lode deposits is in reports and maps by Collier and others (1908), Smith (1910), Moffit (1913), Hummel (1960,

1962a, 1962b), Sainsbury and others (1968), and Sainsbury (1969a, 1969b).

In the Nome area, deposits representative of all major types of placers have been mined. The deposit on Sophie Gulch (16, fig. 28) is a good example of a residual placer, although it has generally been referred to as a lode deposit and was staked as two lode claims and one placer claim. Small scattered iron-stained quartz-feldspar and quartz-calcite veins and schist wallrock contain scheelite and sulfide minerals. Several tons of scheelite concentrates were mined from the weathered material above the lode during World War I and World War II, but the lode itself could not be worked at a profit. Similar deposits on Twin Mountain (14, fig. 28) and Glacier (15, fig. 28) Creeks yielded a total of about half a ton of scheelite. Gold was recovered from residual placers on Pioneer Gulch (22, fig. 28) and Boer Creek (49, fig. 19).

Placers formed mainly by mass-wasting processes have not been described specifically, although such deposits must have been mined in the transition zones between residual and stream placers. Possibly some of the gold in the Caribou Bill claim near Nekula Gulch (30, fig. 28) was moved by mass-wasting processes rather than by stream action.

Stream placers in the gravels of present-day streams, in the gravels of low alluvial terraces (low-bench deposits) along the same streams, and in high channels (the "high benches" of Collier and others (1908, p. 199–209) and Moffit (1913, p. 101–109)), which D. M. Hopkins (oral commun., April 2, 1970) considers to represent glacier-margin channels and spillways, have all been mixed extensively. Some of the stream placers were extremely rich, in part because of reconcentration of material from bench placers, which greatly increased the available supply of gold and other heavy minerals. For example, Snow Gulch (15, fig. 28), a small left-limit tributary of Glacier Creek, is only about three-quarters of a mile long, but gold worth more than a million dollars was taken from it in only a few years. Stream and low bench deposits that were worked during the first 10 years of mining on Anvil Creek probably averaged at least $5 to $6 (gold at $20.67 per ounce) a cubic yard and locally were 10 times as rich. Near the head of Nekula Gulch, a left-limit extreme headwater gulch of Anvil Creek, a hole nearly 90 feet deep (probably a sink or collapsed cavern) in limestone on the Caribou Bill claim was filled with gravel containing coarse angular gold that did not appear to have been carried far but may have been partially reconcentrated from nearby "high-bench" gravels. This material was fabulously rich; its gold content was probably worth about

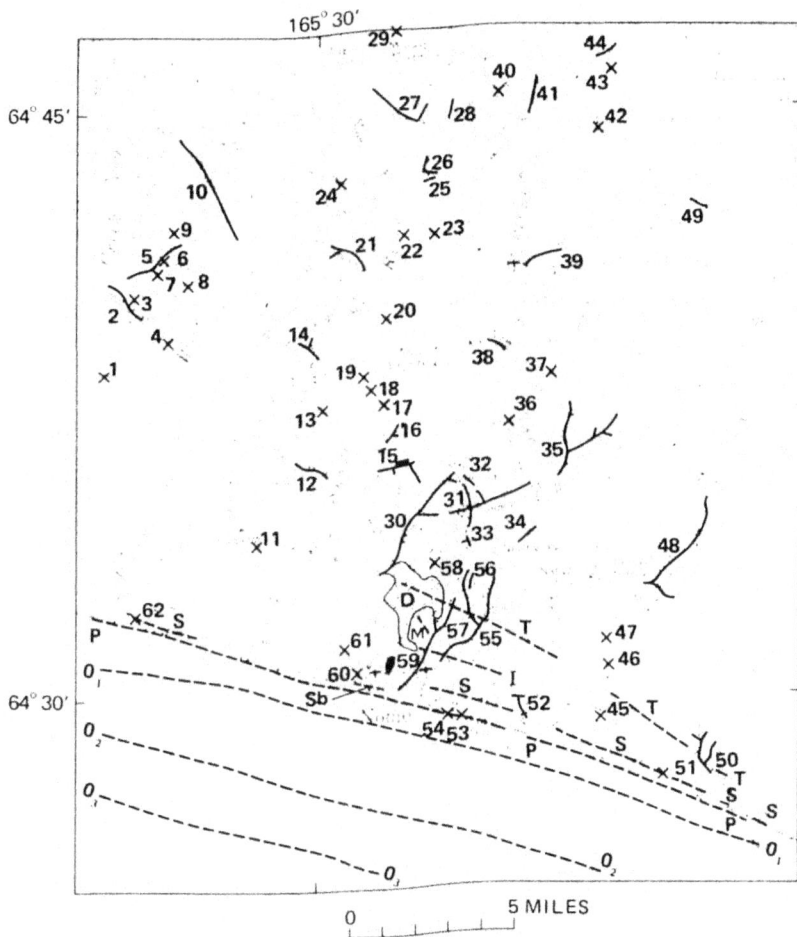

FIGURE 28.—Placer deposits in the Nome area.

1. Arctic Creek: Brooks (1922, p. 63).

2. Hungry Creek: Collier and others (1908, p. 214).

3. Trilby Creek: Collier and others (1908, p. 215).

4. May Gulch: Collier and others (1908, p. 215).

5-6. Oregon Creek: Collier and others (1908, p. 211-213).

7-8. Nugget Creek (Gulch): Collier and others (1908, p. 213-214).

9. Mountain Creek: Collier and others (1908, p. 214).

10. Aurora Creek: Herreid (1968, p. 1-2).

11. Sunset Creek: Smith (1936, p. 49).

12. Monument Creek: Cathcart (1922, p. 190), Anderson (1947, p. 40).
Alpha Creek: Cathcart (1922, p. 249).

14. Boulder Creek: Collier and others (1908, p. 196), Anderson (1947, p. 42). Twin Mountain Creek: Collier and others (1908, p. 197), Cathcart (1922, p. 251).

15. Bonanza Gulch: Moffit (1913, p. 86). Glacier Creek: Moffit (1913, p. 84-85), Mertie (1918b, p. 457), Anderson (1947, p. 40). Hot Air Bench: Moffit (1913, p. 85-86). Snow Gulch: Moffit (1913, p. 85), Coats (1944c, p. 5-6).

16. Rock Creek: Moffit (1913, p. 75-76, 86), Anderson (1947, p. 42). Sophie Gulch: Cathcart (1922, p. 182, 245-246), Coats (1944c, p. 3).

17. Lindblom Creek: Moffit (1913, p. 86-87), Anderson (1947, p. 42).

$1,000 per cubic yard (gold at $20.67 per ounce). Creek and low-bench deposits on some of these and other tributaries of the Snake

---

FIGURE 28.—Continued.

18. Prospect Creek: Thorne and others (1948, p. 33–34).
19. Balto Creek: Moffit (1913, p. 87), Thorne and others (1948, p. 33–34).
20. Divining Creek: Thorne and others (1948, p. 33–34).
21. Bangor Creek: Moffit (1913, p. 87). Butterfield Creek (Canyon): Thorne and others (1948, p. 33).
22. Pioneer Gulch: Moffit (1913, p. 76).
23. Seattle Creek: Coats (1944c, p. 6).
24. Last Chance Creek: Moffit (1913, p. 87).
25. Grub Gulch: Moffit (1913, p. 88).
26. Goldbottom Creek: Moffit (1913, p. 87). Grouse Creek: Moffit (1913, p. 88).
27. Goldbottom Creek: Moffit (1913, p. 87). Steep Creek: Moffit (1913, p. 88).
28. Grouse Creek: Moffit (1913, p. 88).
29. Fred Gulch: Hess (1906, p. 157), Moffit (1913, p. 101).
30. Anvil Creek: Collier and others (1908, p. 191), Moffit (1913, p. 79–83). Nekula Gulch: Moffit (1913, p. 83–84, 101–103). Specimen Gulch: Moffit (1913, p. 84, 106–107).
31. Deer Gulch: Moffit (1913, p. 95, 103). Dexter Creek: Moffit (1913, p. 93–94, 108–109). Dexter Creek, Left Fork: Collier (1905, p. 127), Moffit (1913, p. 95, 107–108). Grass Gulch: Moffit (1913, p. 94–95, 106–108). Grouse Gulch: Moffit (1913, p. 95–96).
32. Dexter Station: Moffit (1913, p. 101–106).
33. Bear Gulch: Brooks (1901, p. 76). Dry Creek: Moffit (1913, p. 90, 101, 107–108).
34. Extra Dry Creek: Moffit (1913, p. 98–99).
35. Buster Creek, Grace Gulch, and Lillian Creek: Moffit (1913, p. 96–97). Union Gulch: Collier and others (1908, p. 173).
36. Nome River: Moffit (1913, p. 93).
37. Dewey Creek: Brooks (1901, p. 78–79).
38. Banner Creek: Moffit (1913, p. 99).
39. Basin Creek: Moffit (1913, p. 99–100).
40. Manila Creek: Chapin (1914b, p. 389).
41. Hobson Creek: Collier and others (1908, p. 181).
42. Darling Creek: Alaska Department of Mines (1948, p. 43).
43. Nelson Creek: Coats (1944c, p. 4–6).
44. Rocky Mountain Creek: Coats (1944c, p. 4–6), Anderson (1947, p. 40, 42).

45. Irene Creek: Mertie (1918b, p. 454), Alaska Department of Mines (1940, p. 85).
46. Moss Gulch: Mertie (1918b, p. 455).
47. Washington Gulch: Moffit (1913, p. 98).
48. Osborn and St. Michaels Creeks: Moffit (1913, p. 97–98).
49. Hazel Creek: Moffit (1913, p. 101).
50. Hastings Creek: Moffit (1913, p. 100–101). Saunders Creek: Chapin (1914b, p. 390).
51. Derby Creek: Mertie (1918b, p. 455).
52. Otter Creek: Chapin (1914b, p. 390).
53. Peluk Creek: Chapin (1914b, p. 389–390).
54. Rocker Gulch: Chapin (1914b, p. 389–390).
55. Dry Creek: Moffit (1913, p. 90). Newton Gulch: Moffit (1913, p. 91–92).
56. Dry Creek bench: Moffit (1913, p. 90–91).
57. Bourbon and Holyoke Creeks: Moffit (1913, p. 89–90).
58. Cooper Gulch: Moffit (1913, p. 89).
59. Dredged area: Alaska Division of Mines and Minerals (1962, p. 8), D. M. Hopkins (oral commun., April 2, 1970).
60-61. Snake River: Moffit (1913, p. 77–79), Smith (1926, p. 18).
62. Jess Creek: Moffit (1906b, p. 133–134).
D. Major dredged area. Center, Flat, Holyoke, Saturday, and Worler Creeks: Moffit (1913, p. 88–90). Lake Creek: Smith (1936, p. 59). Little Creek: Collier and others (1908, p. 170).
I. Intermediate beach: Moffit (1913, p. 117–123).
M. Monroeville beach: Moffit (1913, p. 119–123).
O1. 36-foot beach: Nelson and Hopkins (1969).
O2. 65-foot beach: Nelson and Hopkins (1969).
O3. 75-foot beach: Nelson and Hopkins (1969).
P. Present beach: Brooks (1901, p. 85–91), Moffit (1913, p. 110–111).
S. Second beach: Moffit (1913, p. 40–44, 111–112, 119–123).
Sb. Submarine beach: Moffit (1913, p. 118–123), Nelson and Hopkins (1969).
T. Third beach: Moffit (1913, p. 40–44, 112–117, 119–123), Chapin (1914b, p. 389).

and Nome Rivers were mined, some almost continuously, from the time of their discovery until at least as recently as 1968. Every means of mining from shovel to dredge has been used. Ditches and pipelines, some tapping sources as far away as the south flank of the Kigluaik Mountains, brought the water needed for large-scale operations. The Miocene ditch, the first major ditch constructed on the Seward Peninsula, was begun in 1901 and completed in 1903, paving the way for the integrated operations that made possible large-scale mining in the Nome area.

The "high-bench" placers near Nome were preserved from subsequent erosion in the divide between the headwater gulches of Anvil, Dexter, and Dry Creeks (30–33, fig. 28) in the saddle between King Mountain and the hills to the south (Moffit, 1913, fig. 11). As these deposits were far from any dependable source of water for sluicing, only the richest material (at least $6 per yard) could be mined. Erosion of the "high-bench" placers probably contributed much of the gold in the rich stream placers in Anvil, Dry, and Dexter Creeks.

A diffuse paystreak that Gibson (1911, p. 464–465; fig. on p. 425) considered to be an old channel of Anvil Creek extends from the vicinity of Moonlight Springs, where the stream leaves the mountains, to the coast near the mouth of the Snake River. This deposit, considered by later geologists to be a marine abrasion-platform deposit (D. M. Hopkins, oral commun., April 2, 1970), was drift-mined locally and was later extensively dredged. Enough gold was recovered between 1904 and 1906 from a drift mine on one claim to make the operator a millionaire.

Beach deposits have been a major source of gold in the Nome area since the present beach (P, fig. 28) was found to be auriferous in 1899. As creeks that crossed the tundra between the shore and the base of the hills north of Nome were prospected and mined, other beaches were found. Although not the bonanzas of the richest parts of some of the creeks draining Anvil Mountain, many parts of the beaches were very rich by any other standard. Collier and others (1908, p. 159–165), Gibson (1911), Moffit (1913, p. 109–123), and Hopkins, MacNeil, and Leopold (1960), among others, studied these deposits, recognized their origin, chronicled their exploration, and speculated on their chronology and the events in the complex regional glacial history that allowed their formation and preservation. Interest in finding possible submarine beaches off the coast led to extensive offshore studies by both Federal agencies and private companies.

The following discussion of the beaches and offshore deposits is based mainly on a report by Nelson and Hopkins (1969). Eustatic fluctuations of sea level during Pliocene and Pleistocene time resulted in shifting the strandline as far inland as the base of the hills (Fourth beach) and as far offshore as "75-foot beach" (figs. 28 and 29). A beach deeply buried a short distance offshore and the so-called "Inner" and "Outer Submarine Beaches" are of Pliocene age; the others are probably Pleistocene. Fourth beach is generally too lean to be mined at a profit, although it has contributed gold to minable placers that were formed in streams that cut through it. Plans for prospecting and eventually mining any gold that may be found in the basal parts of the offshore beaches (which could not be sampled with the equipment available to Nelson and Hopkins) are still (1970) in formative stages.

Onshore beaches, including the two Submarine Beaches, were mined in the early 1900's by drifting from shafts sunk from the surface. Dumps were accumulated near the shafts and sluiced with what water was available during the short spring runoff season. This water shortage and the inherent high expense of drift mining soon caused the amalgamation of large blocks of claims. Subsequent large-scale dredging, preceded by cold-water thawing of permafrost in areas to be dredged, allowed many of the remaining old beach deposits and much of the intervening auriferous glacial drift (particularly at 59 and D, fig. 28) to be mined at a profit. Unfavorable economic factors rather than exhaustion of resources finally caused large-scale operations to cease at the end of the 1962 season. Production from the old beaches accounted for most of the gold recovered in the Nome area and possibly for as much as half of that reported for the entire Seward Peninsula region.

Thin auriferous deposits not associated with old beach lines were recently found on the floor of the Bering Sea near Nome. According to Nelson and Hopkins (1969), these are in places where wave action during shoreline transgressions and regressions has winnowed fine material from glacial drift, leaving relict gravel resting on relatively unsorted till, outwash, and alluvium. Some samples of this material contained as much as $4 in gold per cubic yard, and about one-third of those recovered during the investigations of Nelson and Hopkins contained gold worth more than $1 per cubic yard (gold at $35.00 per fine ounce). The thinness of the relict gravel, which averages about 1 foot thick, could present serious technological problems for economic recovery of its gold. Gold in generally similar gravel on the sea bottom near Sledge Island about 25 miles west of Nome was probably derived from nearby bedrock sources that are now submerged, rather than from lode deposits on what is now the mainland.

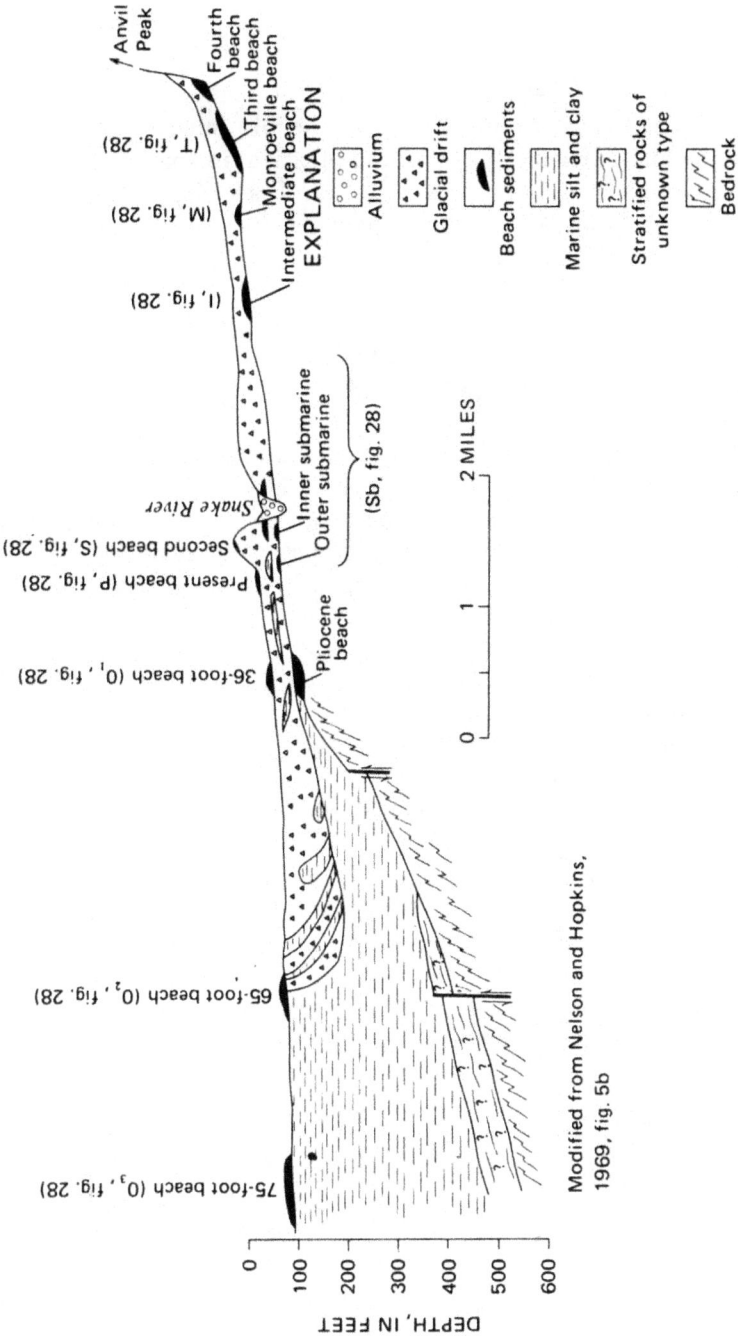

FIGURE 29.—Generalized geologic profile of Nome beaches.

Sesimic profiling offshore from Nome located what are probably old buried stream channels that could be inviting exploration targets (A. R. Tagg, oral commun., April 2, 1970).

The commonest heavy minerals that accompany gold in placer concentrates in the Nome area are scheelite, magnetite, ilmenite, hematite, and garnet. Scheelite is so plentiful in some deposits that it has been produced from Snow Gulch, Rock Creek (16, fig. 28), and probably from a few other streams. Cassiterite has been reported from the Left Fork of Dexter Creek (31, fig. 28) and from Glacier, Goldbottom (26, fig. 28), Monument (12, fig. 28), and Rocky Mountain (44, fig. 28) Creeks but not in sufficient quantities to be worth saving. Stibnite and bismuthinite have been found at Boulder Creek (14, fig. 28) and stibnite has been found at Dorothy Creek (53, fig. 19). Submarine beach contains the largest amounts of pyrite, chalcopyrite, and, in particular, arsenopyrite of any of the beach gravels near Nome or creek placers in the Snake River and Nome River drainage basins. The presence of these relatively "fragile" heavy minerals suggests an undiscovered nearby bedrock source that might also have contributed gold.

East of the Nome River, at the point where Hazel Creek (49, fig. 28) flows from the hills into the broad Flambeau River valley, there was placer mining in at least one year. A few miles beyond the divide west of the Snake River in areas geologically similar to areas closer to Nome, mining has been carried on in the basins of the Sinuk and Cripple Rivers. On the basis of incomplete data, most of the mining was apparently on Oregon (5, 6, fig. 28) and Arctic (1, fig. 28) Creeks. Coal Creek (45, fig. 19), though very little gold was produced there, is interesting in that at least some of the gold has been reconcentrated from Tertiary coal-bearing rocks. Native bismuth and rutile have been identified in concentrates from Charley Creek (47, fig. 19) and several other streams. Scheelite has been reported from Oregon and Nugget (7, 8, fig. 28) Creeks, and cassiterite has been reported from Fred Gulch (29, fig. 28). Near Aurora Creek (10, fig. 28) widely distributed float carries a considerable amount of sphalerite and a little galena and chalcopyrite.

The Solomon area (fig. 30) is geologically similar to the Casadepaga area of the Council district and to the rest of the Nome district. A lode deposit near Solomon—the Big Hurrah mine near the head of Big Hurrah Creek (13, fig. 30)—was rich enough to be mined and yielded at least 10,000 fine ounces of gold between 1900 and World War 11 (Berg and Cobb, 1967, p. 126–127). This and a neighboring prospect contain enough scheelite to have been investigated as sources of tungsten. Several other lodes in the

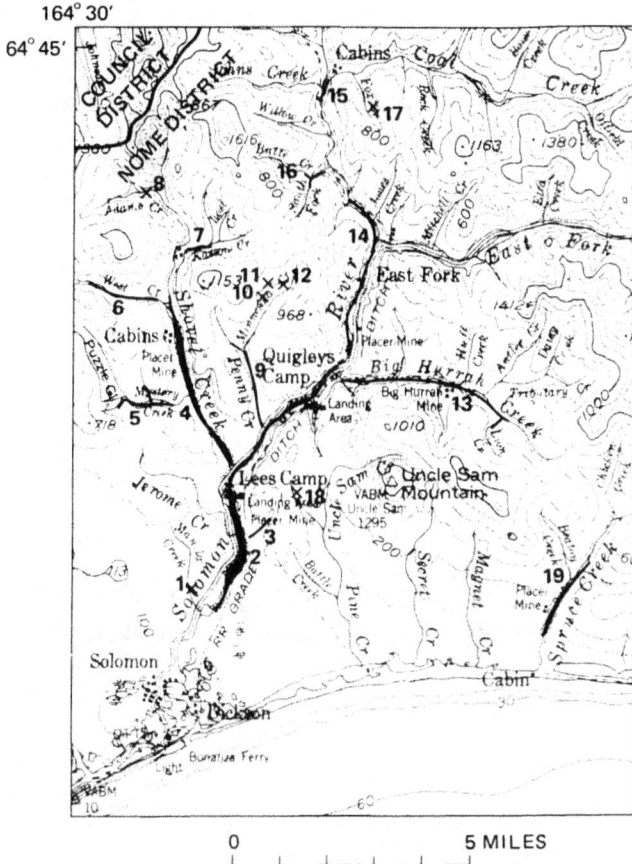

FIGURE 30.—Placer deposits in the Solomon area.

1. Manila Creek: Smith (1910, p. 170).
2. Jerome Creek: Smith (1910, p. 171). Moran Gulch: Smith (1910, p. 161). Quartz Creek: Smith (1910, p. 180). Solomon River: Smith (1910, p. 155–168), Smith (1942b, p. 62), Coats (1944a, p. 4).
3. Rock Creek: Smith (1910, p. 170–171).
4. Shovel Creek: Smith (1910, p. 171–173).
5. Mystery Creek: Smith (1910, p. 172–174). Problem and Puzzle Gulches: Smith (1910, p. 173, 175).
6. West Creek: Smith (1910, p. 175–177).
7. Kasson Creek: Smith (1910, p. 177–178).
8. Adams Creek: Smith (1910, p. 178).
9. Penny Creek: Smith (1910, p. 179).
10. Sapphire Gulch: Smith (1910, p. 179).
11. Meddler Gulch: Smith (1910, p. 179–180).
12. Bear Gulch: Smith (1910, p. 179).
13. Big Hurrah Creek: Smith (1910, p. 180–182), Coats (1944a, p. 3–4). Lion Creek: Collier and others (1908, p. 232).
14. Solomon River: Smith (1910, p. 168–169).
15. Solomon River: Smith (1910, p. 169), Smith (1933a, p. 47, 54).
16. Butte Creek and South Fork: Smith (1910, p. 183).
17. Fox Creek: Smith (1910, p. 184).
18. Rabbit Creek: Smith (1910, p. 212–213).
19. Spruce Creek: Smith (1910, p. 203–204), Smith (1939a, p. 69–70).

area were prospected for gold or antimony, but none was ever brought into production.

Placer mining in the Solomon River valley dates from 1899, when the first claim was staked on the Solomon River near the mouth of Big Hurrah Creek (2, fig. 30). Stream and bench placers were worked until 1967, when activity had dwindled to two two-man operations on Shovel Creek (4, fig. 30). Most of the gold mined was recovered by dredges from the Solomon River (2, 14, 15, fig. 30), its major tributaries, Shovel and Big Hurrah Creeks, and Spruce Creek (19, fig. 30), to which a dredge was moved from Shovel Creek in 1929. The richest gravels were probably those of Big Hurrah Creek below the Big Hurrah mine. The last dredge to operate in the area was dismantled about 1963 and the machinery was moved from the Solomon River to the western end of the Seward Peninsula.

Stream and bench deposits in the Solomon area were mined by methods other than dredging, particularly on Kasson Creek (7, fig. 30), on Mystery Creek and its tributaries Problem and Puzzle Gulches (5, fig. 30), on West Creek (6, fig. 30), in the Shovel Creek drainage basin, and on Penny Creek (9, fig. 30).

Scheelite is common in concentrates from Big Hurrah Creek and the Solomon River. Coats (1944a, p. 4) noted that analysis of a sample of a dredge concentrate from the Solomon River below the mouth of Shovel Creek indicated 22 ounces of gold and 9.1 pounds of scheelite per cubic yard of concentrate. There is no record that tungsten concentrates from the Solomon area were ever marketed. Other heavy minerals reported include magnetite, ilmenite, garnet, pyrite, chalcopyrite, and arsenopyrite.

There has been some mining on at least three west-flowing tributaries of the Eldorado River, the largest stream between the Nome and Solomon Rivers. In the 1940's parts of Beaver and Pajara Creeks (54, fig. 19) were dredged. Between 1900 and 1903, gold worth a few thousand dollars was recovered from Venetia Creek with less elaborate equipment (55, fig. 19).

PORT CLARENCE DISTRICT

The Port Clarence district (fig. 19) includes the area drained by streams flowing into the Arctic Ocean and the Bering Sea between Shishmaref Inlet and Cape Douglas and, with the exception of the Kuzitrin River, by streams flowing into Imuruk Basin.

Lode deposits in the district have been investigated as possible sources of tin, tungsten, beryllium, gold, silver, lead, zinc, copper, antimony, bismuth, molybdenum, uranium, graphite, and fluorite (Sainsbury, 1969a; Berg and Cobb, 1967, p. 128–135, fig. 23). Lodes in Lost River and Cape Mountain have been the sources of all the lode tin produced in Alaska, a total of about 250 tons. A few tons of graphite has been mined from graphitic schist in the north flank of the Kigluaik Mountains near Imuruk Basin (Coats, 1944b).

In 1898, the first prospectors in the Port Clarence district failed to locate workable placers, but in the next year, gold was found on Buhner (Buckner) Creek (15, fig. 32) and the Bluestone River (11–13, fig. 31) and in 1900, cassiterite (stream tin) was identified in Buhner Creek and the Anikovik River (13, fig. 32) (Collier and others, 1908, p. 269, 273; Knopf, 1908, p. 7). Mining for both tin and gold have continued to the present (1970). Total tin production from the Seward Peninsula through 1954 was 2,200 tons of tin metal (Sainsbury, 1969a, p. 62), most of which came from placer deposits in streams that drain Potato and Cape Mountains (fig. 32). Data on gold production in the Port Clarence district, mainly from streams in the Teller area (fig. 31), are not available, but the amount was probably much less than from any other district in the Seward Peninsula region except the Serpentine district.

Geologically the Port Clarence district is generally similar to the northern part of the Kougarok district. Bedrock is mainly complexly faulted slate, schist, and limestone, gabbroic sills, felsic and mafic dikes, and small granitic plutons (Sainsbury, 1969a, b, c). Most of the tin placers are on streams that drain contact zones around tin-bearing granites or dikes probably associated with as yet unexposed plutons.

Data on mining in the eastern part of the Port Clarence district are incomplete and confusing, partly because several streams have been given the same name and in some instances various names or variant spellings were given a single feature or place. For example, there are at least three streams, two of which were sites of placer mining, that were called Igloo Creek at one time or another. Two Windy Creeks have been listed as producing streams, and both Gold Run and Goldrun Creek have been mined for many years. The name of Allene Creek has been changed several times.

South of Grantley Harbor most of the gold recovered has come from creek and bench (some as much as 200 feet above the present streams) deposits along Gold Run (13, fig. 31), Coyote (9, fig. 31), and Dese (10, fig. 31) Creeks and right-limit tributaries of the Right Fork of the Bluestone River (16–19, fig. 31). Dredges or other mechanized equipment in use before 1942 accounted for most of the production, although smaller operations were reported for most years since gold mining was resumed after World War II.

Within a few miles of tidewater north of Grantly Harbor, Sun- (1, fig. 31), Allene (6–8, fig. 31), and Offield (5, fig. 31)

Creeks have been extensively mined. Concentrates from Sunset Creek contained so much scheelite that in 1917, a time of high tungsten prices, it was worth saving and was marketed. The longest ditch in the Port Clarence district was constructed to bring water from the headwaters of the Agiapuk River to Sunset Creek. Farther north, much small-scale mining has been reported from tributaries of the American River, which rises about 25 miles southwest of Kougarok Mountain and flows into the Agiapuk River. Many of these operations were on small streams that bore names not shown on available maps. The only major production from the American River drainage basin was from near the junction of Windy and Budd Creeks (60, fig. 19).

Data on heavy minerals in concentrates from streams in the eastern part of the Port Clarence district are scarce, although the recovery of a loaded revolver from a sluicebox on a bench of Gold Run was reported! In addition to the scheelite mined at Sunset Creek, cinnabar and copper minerals, both from Budd Creek, have been identified. There are unsubstantiated reports of cassiterite at Budd and Windy Creeks.

In 1920, about 20 tons of garnet sand from a beach of Imuruk Basin was shipped from Nome. The exact location of this beach is not known, but it is probably somewhere near locality 20 of figure 31.

The western part of the Port Clarence district, originally prospected for gold, is essentially a tin province. After cassiterite was first identified in samples from Buhner Creek (15, fig. 32) the entire area was prospected for both lode and placer tin. Most of the small amount of placer gold recovered was incidental to tin mining. In addition to considerable exploration by private companies, both the U.S. Bureau of Mines and the Geological Survey have carried on extensive investigations of both lode and placer tin resources. Most of the work was done near Cape Mountain at the westernmost extremity of the Seward Peninsula, at Potato Mountain about 15 miles to the east, at Lost River (57, fig. 19), and at Ear Mountain southwest of Shishmaref Inlet. Each of these areas, except Ear Mountain, has contributed a significant part of the more than 2,200 tons of tin metal derived from lodes and placers in the district.

The most productive group of tin placers was near Potato Mountain on Buck and Grouse Creeks (20, fig. 32), where a dredge was operated for many years and nonfloat mining continued until 1954. No large body of tin-bearing granite is exposed at Potato Mountain, but dikes suggest that there may be one at depth. Some

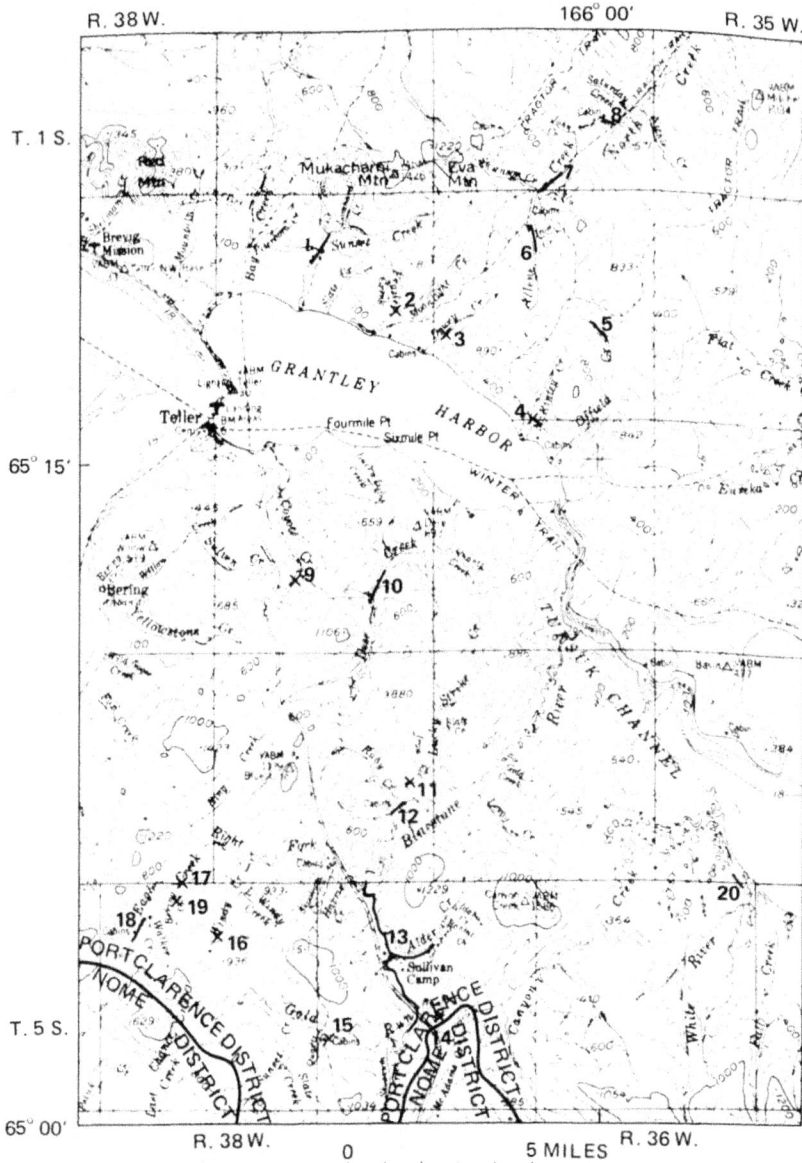

FIGURE 31.—Placer deposits in the Teller area.

1. Sunset Creek: Martin (1919, p. 41), White, West, and Matzko (1953, p. 2), Sainsbury (1967, p. D210).
2. Igloo (Moonlight) Creek: Collier and others (1908, p. 270–271).
3. Dewey Creek: Collier and others (1908, p. 270–271).
4. McKinley Creek: Collier and others (1908, p. 270–271).

5. Offield Creek: White, West, and Matzko (1953, p. 2).
6-8. Allene (Ilene, Swanson) Creek: Collier and others (1908, p. 271–272), White, West, and Matzko (1953, p. 2), Cobb and Sainsbury (1968, p. 5).
9. Coyote Creek: Smith (1942b, p. 63).
10. Dese Creek: Smith (1938, p. 64), White, West, and Matzko (1953, p. 2).

cassiterite has been mined from Sutter (20, fig. 32) and Iron (19, fig. 32) Creeks in the Buck Creek drainage basin and cassiterite has been found on the other side of Potato Mountain in Potato (17, fig. 32) and Diomede (Oakland) (18, fig. 32) Creeks. A little gold accompanies the cassiterite in Buck, Grouse, and Sutter Creeks, and various other heavy minerals—including magnetite, hematite, rutile, and possibly monazite and scheelite—have been reported from concentrate samples from Buck Creek.

Cape (Tin City) Creek and its tributary First Chance Creek (2, fig. 32) and Goodwin Creek and Goodwin Gulch (3, fig. 32) drain the eastern slope of Cape Mountain, which is underlain by a granite stock in which tin-bearing lode deposits have long been known. Goodwin Gulch or Cape Creek, or both, have been mined in most years since the early 1920's. Cape Creek is currently (1970) being worked with machinery from a dredge formerly used on the Solomon River. Results are rumored to be better than had been anticipated. Cassiterite was found on Village (1, fig. 32) and Boulder (4, fig. 32) Creeks, which drain the north flank of Cape Mountain, but no mining has been reported on either of them. In addition to cassiterite, concentrates from streams in the Cape Mountain area contain scheelite, monazite, xenotime (YPO_4), and zircon. Columbium (niobium) and tantalum have been identified by spectrographic analysis.

In the Lost River area (56–57, fig. 19) (Sainsbury, 1964), lode deposits of the Lost River mine have attracted most of the interest. The mine has been the source of about 350 tons of tin metal. Placer deposits, derived from the cassiterite-bearing lodes, have been mined intermittently on Cassiterite Creek (57, fig. 19) and Lost River. Between 1949 and 1951, concentrates containing about 93.4 tons of tin metal were produced. Although placer mining on Lost River was reported in 1966 and 1967, no figures on the results of these operations have been published. Wolframite has been found, but not saved, in concentrates from Cassiterite Creek.

FIGURE 31.—Continued.

11. Bluestone River: Collier and others (1908, p. 273 275).

12. Bluestone River: Cobb and Sainsbury (1968, p. 5).

13. Alder Creek: Collier and others (1908, p. 279 280). Bluestone River: Collier and others (1908, p. 276-278). Gold Run: Collier and others (1908, p. 275-279), Anderson (1947, p. 43-44), White, West, and Matzko (1953, p. 1).

14. Gold Run: Collier and others (1908, p. 279).

15. Gold Run: Collier and others (1908, p. 277).

16. Windy Creek: Smith (1933b, p. 50).

17. Igloo (Eagle) Creek: White, West, and Matzko (1953, p. 1).

18. Igloo (Eagle) Creek: Cobb and Sainsbury (1968, p. 5).

19. Bering Creek: Collier and others (1908, p. 280 281).

20. Imuruk Basin: Brooks (1922, p. 33). (Location approximate only.)

The Ear Mountain area (fig. 33), which includes parts of both the Port Clarence and Serpentine districts, is underlain by Paleo-

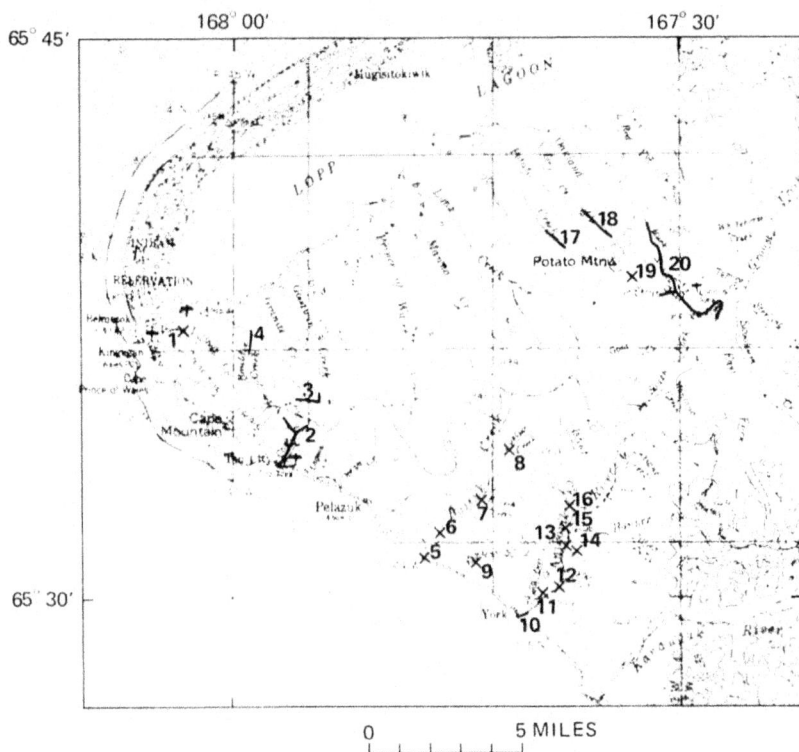

FIGURE 32.—Placer deposits in the Cape Mountain-Potato Mountain area.

1. Village Creek: Heide and Sanford (1948, p. 4, 6, 13).

2. Cape (Tin City) Creek: Mulligan and Thorne (1959, p. 20, 43, 45-47), Barton (1962, p. 31), Mulligan (1966, p. 18, 20-23, 29), Alaska Division of Mines and Minerals (1966 p. 11, 104). First Chance Creek: Mulligan (1966, p. 18, 20-21, 23).

3. Goodwin Creek: Mulligan (1966, p. 18-19, 21). Goodwin Gulch: Mulligan (1966, p. 18-19, 23, 29).

4. Boulder Creek: Mulligan and Thorne (1959, p. 47-66), Mulligan (1966, p. 18-19, 21).

5 7. Baituk Creek: Mulligan (1959b, p. 21-23).

8. Baituk (Justice) Creek: Mulligan (1959b, p. 21-23).

9. Kigezruk Creek: Brooks (1901, p. 135), Mulligan (1959b, p. 21).

10. Anikovik River: Mulligan (1959b, p. 5).

11. Deer Creek: Brooks (1901, p. 134-135).

12. Anikovik River: Mulligan (1959b, p. 15, 17-20).

13. Anikovik River: Brooks (1901, p. 136-137).

14. Banner Creek: Brooks (1901, p. 135), Mulligan (1959b, p. 19).

15. Buhner (Buckner) Creek: Brooks (1901, p. 135-136).

16. Ishut Creek: Brooks (1901, p. 135), Mulligan (1959b, p. 19).

17. Potato Creek: Heide and Rutledge (1949, p. 4, 7, 19).

18. Diomede (Oakland) Creek: Heide and Rutledge (1949, p. 7, 20).

19. Iron Creek: Heide and Rutledge (1949, p. 7-8, 15).

20. Buck Creek: Steidtmann and Cathcart (1922, p. 94-96), Mulligan (1965a, p. 23-31, 62-64). Grouse Creek: Eakin (1915a, p. 91), Mulligan (1965a, p. 9-11, 24-25). Peluk Creek: Mulligan (1965a, p. 24, 27-30). Sutter Creek: Mulligan (1965a, p. 9-11, 24-25, 56-57).

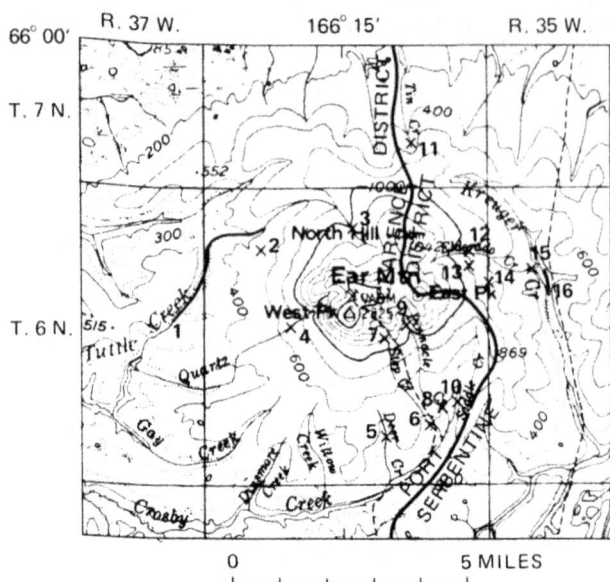

FIGURE 33.—Placer deposits in the Ear Mountain area.

1. Tuttle Creek: Killeen and Ordway (1955, p. 69, 82), Mulligan (1959a, 21–22, 33).
2. Tuttle Creek: Mulligan (1959a, p. 31).
3. Tuttle Creek: Mulligan (1959a, p. 30, 32–33).
4. Quartz Creek: Killeen and Ordway (1955, p. 82), Mulligan (1959a, p. 30–31).
5. Deer Creek: Mulligan (1959a, p 29–30, 32).
6. Step (Gulch) Creek: Mulligan (1959a, p. 29–30, 32).
7. Step Gulch: Killeen and Ordway (1955,

p. 71, 79, 81, 83).
8. Pinnacle Creek: Mulligan (1959a, p. 29–30, 32).
9. Pinnacle Creek: Killeen and Ordway (1955, p. 82–83).
10. Crosby Creek: Mulligan (1959a, p. 29–30, 32).
11. Tin Creek: Mulligan (1959a, p. 30, 32).
12–14. Eldorado Creek: Mulligan (1959a, p. 1–3, 24, 29–30, 32–33).
15. Eldorado Creek: Killeen and Ordway (1955, p. 82).
16. Kreuger Creek: Mulligan (1959a, p. 24, 33).

zoic schist and limestone and a granite stock and associated dikes. The mineralized contact aureole between the bedded and intrusive rocks was the source of the heavy minerals in the gravels of streams draining Ear Mountain. Cassiterite was discovered on Eldorado Creek (12–15, fig. 33) in 1901, but not in minable amounts. As a result of finding high radioactivity in old samples in collections of the Geological Survey, Killeen and Ordway (1955) visited the Ear Mountain area in 1945 and carried out an extensive sampling program. They found cassiterite to be ubiquitous, but not in minable amounts. They also identified monazite, zircon, xenotime, and scheelite and found a little gold in Tuttle Creek (1, fig. 33). Columbium (niobium) was identified by spectrographic methods in samples from Tuttle and Quartz (4, fig. 33) Creeks during a subsequent investigation by the U.S. Bureau of Mines (Mulligan, 1959a).

Several streams that drain into the Bering Sea between Lost Rver and Cape Creek contain cassiterite. The Anikovik River and its tributaries Deer, Banner, Buhner, and Ishut Creeks (10–16, fig. 32) were prospected for gold in 1900. A little gold was found in all of them, but there was not enough to mine except on the Anikovik, where between 1,225 and 1,250 ounces was recovered within about half a mile of the coast (10, fig. 32). Most of the gold came from a dredge operation in 1914 and 1915 that also resulted in the production of about 1,600 pounds of concentrate containing 31 percent tin. Chromite was identified in a churn-drill sample a short distance upstream (12, fig. 32) and scheelite was identified in a similar sample from Ishut Creek (16, fig. 32).

Mulligan (1959b, p. 15–17) reported small amounts of scheelite, powellite, and cassiterite in churn-drill samples from the headwaters of York Creek (58, fig. 19) about 2 miles north of a granite stock exposed at Brooks Mountain. According to Anderson (1947, p. 44), wolframite has been reported from the upper Pinguk River 15–20 miles east of Brooks Mountain, but this occurrence has not been confirmed.

SERPENTINE DISTRICT

The Serpentine district (fig. 19) includes the area drained by streams flowing into the Arctic Ocean and Kotzebue Sound from (and including) Pish River to Shishmaref Inlet.

Most of the district is underlain by surficial deposits and volcanic rocks of late Cenozoic age (Hopkins, 1963, pl. 2). The southern fringe is a continuation of the northern Kougarok and northwestern Fairhaven districts. The single known lode in the Serpentine district is a copper deposit a few miles north of Kougarok Mountain in limestone near a contact with mica schist (Berg and Cobb, 1967, p. 135, fig. 23).

The only commercial placer mining in the district was along Dick Creek (61, fig. 19), which heads against the northern part of the Kougarok River drainage. Operations were carried on there for many years, but they ceased at the end of the 1952 season, when the deposit was worked out. Scheelite and cassiterite were found with gold in the concentrates. Other streams in the same general area were prospected about 1901 (Collier, 1902, p. 55), but evidently the results were not encouraging, as later reports did not note production from any of them.

Placer tin occurrences near Ear Mountain (fig. 33) were discussed with descriptions of neighboring creeks in the Port Clarence district.

In 1946, Moxham and West (1953) investigated the headwater area of Hot Springs Creek (62, fig. 19) for radioactive minerals. They found small amounts of allanite and a little cinnabar. Although Hot Springs Creek heads in the same area as Humboldt Creek in the Fairhaven district, no gold or cassiterite was reported.

SOUTHEASTERN ALASKA REGION

The southeastern Alaska region (pl. 1, fig. 34) is the part of Alaska east of long 141° W. The Admiralty, Chichagof, Hyder, Juneau, Ketchikan, Kupreanof, Petersburg, and Yakutat districts are in this region.

The region is essentially a mountainous area composed of a wide belt of rugged islands with summits that rise from 2,500 to 3,500 feet in elevation and a deeply embayed mainland strip whose high peaks rise to over 10,000 feet along the Alaska-Canada boundary. Many of the interisland waterways and major fiords and streams occupy long linear depressions. In the northwestern part of the district, a narrow coastal plain, generally less than 200 feet above sea level and in part covered by the Malaspina Glacier, separates the mountains and the Gulf of Alaska.

Southeastern Alaska is underlain by sedimentary, volcanic, metamorphic, and plutonic rocks that range in age from Paleozoic to Tertiary (Brew and others, 1966). The largest intrusive body is the generally felsic, composite Coast Range batholith exposed along the international boundary for most of the length of the region. The structural history includes several episodes of folding, faulting, intrusion, and metamorphism.

Extensive Pleistocene glaciation greatly modified the preexisting surface of southeastern Alaska. Valleys and interisland waterways were scoured and deepened and most preglacial unconsolidated deposits were removed. Remnants of Pleistocene glaciers and icecaps still cover large areas in the high country adjoining the international boundary, and glacial deposits mantle the lower slopes. In general, the region is free of permafrost.

Gold, silver, copper, lead, zinc, platinum-group metals, tungsten, and uranium have been mined commercially from lodes. Exploratory work, some intensive, has been done at prospects potentially valuable for these metals and for nickel, cobalt, chromite, iron, molybdenum, antimony, and rare-earth elements (Berg and Cobb, 1967, p. 136–195, figs. 24–28). Most of the lodes are in metamor-

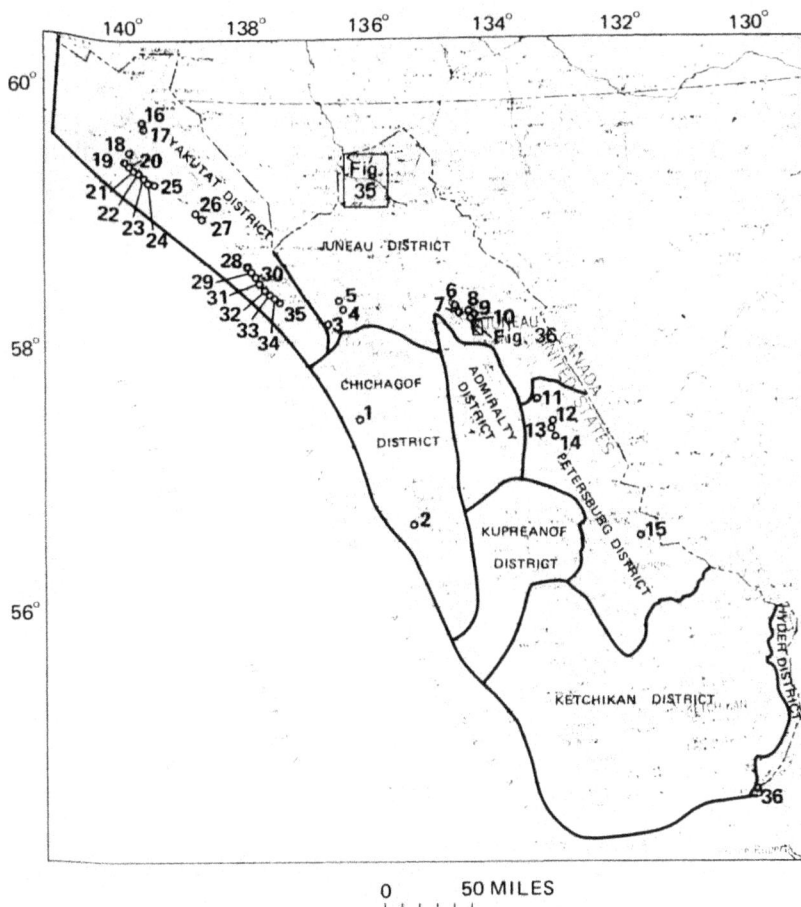

FIGURE 34.—Placer deposits and mining districts in the southeastern Alaska region.

Admiralty district
No placer occurrences.

Chicagof district
1. McKallick: Reed and Coats (1942, p. 124–125).
2. Goddard Hot Springs: West and Benson (1955, p. 47–49).

Hyder district
36. Kanagunut Island: Sainsbury (1957, p. 152).

Juneau district
3. Brady Glacier: Rossman (1963, p. K50–K51).
4. Dundas River: MacKevett and others (1967, p. 120).
5. Dundas River, near Wood Lake: Rossman (1963, p. K50).
6. Windfall Creek: Spencer (1906, p. 127–

Juneau district—Continued
128), Wright and Wright (1906, p. 36).
7. McGinnis Creek: Spencer (1906, p. 123–124). Montana Creek: Spencer (1906, p. 124), Smith (1942b, p. 32).
8. Nugget Creek: Spencer (1906, p. 120–121).
9–10. Lemon Creek: Spencer (1906, p. 119–120).

Ketchikan district
No placer occurrences.

Kupreanof district
No placer occurrences.

Petersburg district
11. Powers Creek: Spencer (1906, p. 2, 45).
12. Spruce Creek: Spencer (1906, p. 42), Alaska Department of Mines (1950, p. 6).

phic rocks adjacent to the Coast Range batholith and satellitic plutons; a few are wholly within the plutons, mainly in marginal facies. Lode production amounted to 7½ to 8 million ounces of gold, several million ounces of silver, about 30 million pounds of copper, more than a quarter of a million pounds of lead, and small amounts of platinum and palladium, tungsten, zinc, and uranium.

Only gold and a little platinum have been recorded from placer deposits in southeastern Alaska. Production records are not complete and in many instances include gold recovered from the neighboring Yakataga district. Probably 120,000–125,000 ounces of gold, or less than 2 percent of the amount from lode sources in the region, came from placers in southeastern Alaska.

Intense alpine glaciation removed most placers that might have been formed as a result of preglacial erosion of lodes. Because most streams have extremely steep gradients and discharge into deep fiords, few gravel deposits are being formed today. Recently, private interests have been investigating sea-bottom tracts, in particular in the Juneau area, for possible submarine placers but have not made public their results.

Most of the placer gold from southeastern Alaska has come from two areas in the Juneau district (fig. 34)—one area is centered near Porcupine Creek (fig. 35) northwest of Haines, the other is close to the city of Juneau (fig. 36). Bedrock in the Porcupine Creek area consists of Paleozoic limestone and clastic rocks that were intruded and metamorphosed by the Coast Range batholith and smaller satellitic plutons. Locally, quartz and calcite stringers and veins carry sulfide minerals and a little fine gold (Wright, 1904b, p. 17–18). Workable placers were found in gravels of the present streams, in low bench deposits along them, and in high bench gravels that are remnants of old valley fills (Wright, 1904b, p. 19; Eakin, 1919, p. 20–21). Steep gradients and abundant water made hydraulic mining the most practical method for working the placers, although floods that filled cuts and destroyed equipment were a constant hazard. Gold was discovered on Porcupine Creek (3, fig. 35) in 1898 and has been mined intermittently since then; in 1968 three one-man opera-

---

FIGURE 34.—Continued.

Petersburg district—Continued

13. Slate and Sylva (Sylvia) Creeks: Spencer (1906, p. 42).

14. Shuck (Chuck) River: Spencer (1906, p. 42-43).

15. Stikine River: Blake (1868, p. 10).

Yakutat district

16-25. Yakutat beaches: Tarr and Butler (1909,

Yakutat district—Continued

p. 165-167), Thomas and Berryhill (1962, p. 26-30).

26-27. Dry Bay: Thomas and Berryhill (1962, p. 32-33, 35-36).

28-35. Lituya Bay: Mertie (1933, p. 133-136), Rossman (1957), and Thomas and Berryhill (1962, p. 37-39).

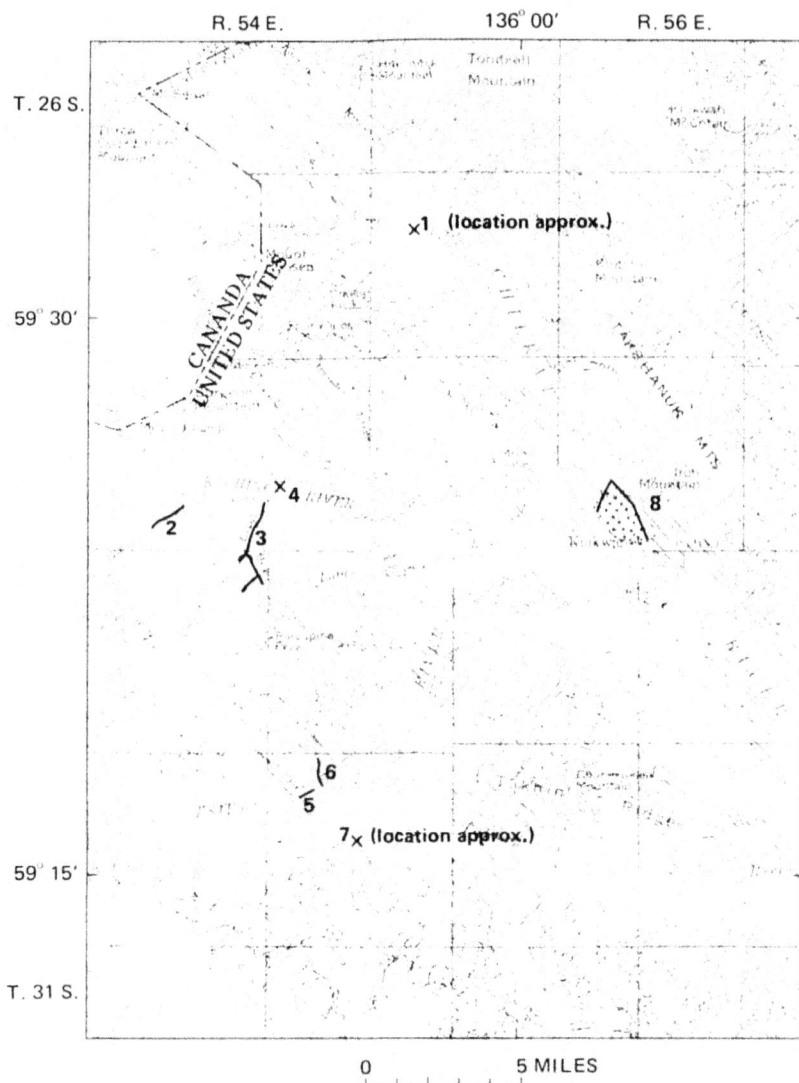

FIGURE 35.—Placer deposits in the Porcupine Creek area.

1. Clear (Rosaunt) Creek: Wright (1904b, p. 13). Kelsall River (Bear Creek): Wright (1904b, p. 63), Eakin (1919, p. 19).
2. Glacier Creek: Eakin (1919, p. 21–23).
3. Cahoon Creek: Eakin (1919, p. 24–25). McKinley Creek: Wright (1904b, p. 24–26), Eakin (1919, p. 21–25). Porcupine Creek: Wright (1904b, p. 19–24), Eakin (1919, p. 22–25).
4. Klehini River: Wright (1904b, p. 22), Smith (1932, p. 27).
5. Cottonwood Creek: Wright (1904b, p. 13). Tsirku (Salmon) River: Wright (1904a, p. 63).
6. Nugget Creek: Wright (1904b, p. 26), Eakin (1919, p. 23).
7. Takhin River: Eakin (1919, p. 23).
8. Klukwan: Robertson (1956, p. 28–36).

tions were reported. Other productive streams were McKinley and Cahoon Creeks (3, fig. 35), both of which are in the Porcupine Creek drainage basin, and Glacier (2, fig. 35) and Nugget (6, fig. 35) Creeks. Gold was found on other creeks, but the deposits were not minable. In addition to gold, concentrates from Porcupine Creek contained galena, magnetite, chalcopyrite, much pyrite, and some arsenopyrite.

An alluvial fan made up of material eroded from a magnetite-bearing pyroxenite body in the Takshanuk Mountains near Klukwan (8, fig. 35) has been extensively investigated by Federal agencies and several private companies since World War II. The fan contains an estimated 500 million tons of broken material with an average magnetite content of 10 percent. A few tons has been taken out for metallurgical testing, but none of the material has been mined commercially.

In the area near the city of Juneau (fig. 36), in 1880, Joe Juneau and Richard Harris found rich gravel and fragments of quartz containing fine gold. They followed these deposits up Gold Creek to the quartz-slate lodes later developed into the famous Alaska-Juneau gold mine. The placers, all in the drainage basin of Gold Creek, were classified by the early miners as hill, gulch, and creek placers. Those called "hill" (mainly near 5, fig. 36) were essentially broken-up lode material. Gulch placers were derived from the lode material by mass wasting on steep slopes that fed into streams, where the lode material was incorporated into leaner stream gravels. Most mining consisted of removing gravel from bedrock basins in the valley of Gold Creek through tunnels driven in bedrock. Silverbow (2, fig. 36) and Last Chance (1, fig. 36) Basins accounted for most of the production. There has been no recent placer mining on creeks near Juneau, but a little gold has been recovered by sluicing tailings from the old Alaska-Juneau mill (12, fig. 36).

Deposits similar to those on Gold Creek were mined on a small scale on several streams between Juneau and Eagle River, about 20 miles northwest of the city, in the late 1800's and early 1900's. Most of these operations were on Windfall (6, fig. 34), Montana (7, fig. 34), and McGinnis (7, fig. 34) Creeks.

Near Windham Bay, an area geologically similar to that near Juneau, placer gold has been recovered from Spruce Creek (12, fig. 34), the Shuck (Chuck) River (14, fig. 34) and some of its tributaries, and Powers Creek (11, fig. 34). Mining was carried on sporadically from as early as 1888 until as recently as 1950, but the amount produced, though not precisely known, was un-

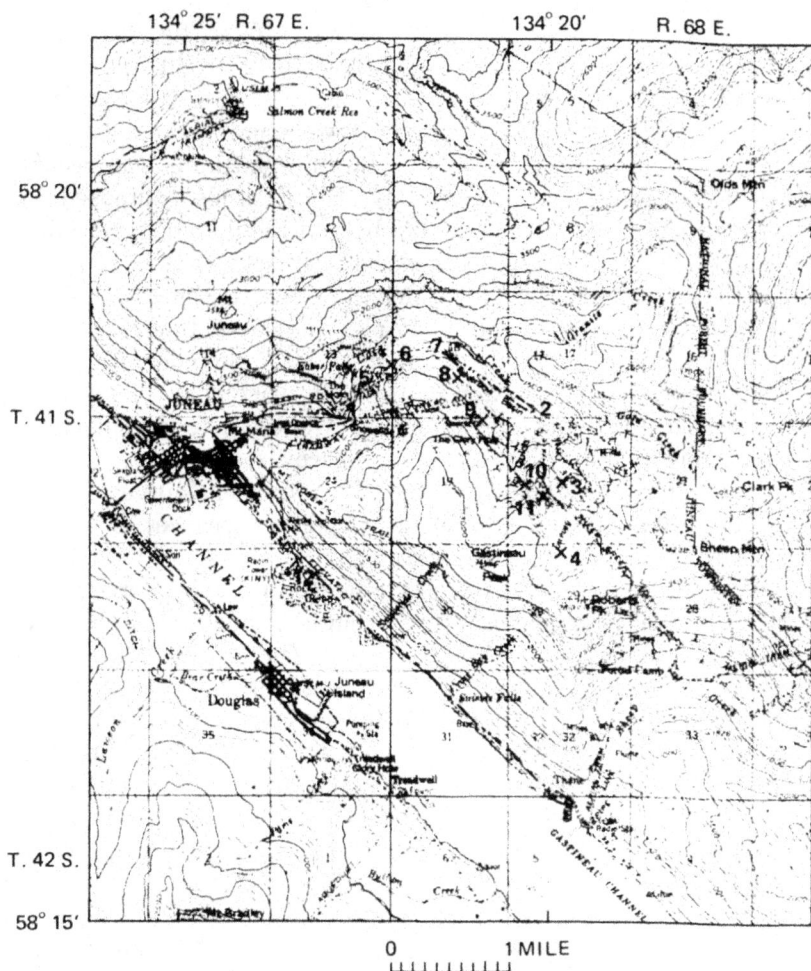

'FIGURE 36.—Placer deposits in the Juneau area. Deposits 1-11, Spencer
    (1906, p. 77-84) ; 12, Berg and Cobb (1967, p. 155).

1. Last Chance Basin (Jualpa).
2. Silverbow Basin (Nowell).
3. Lurvey.
4. Lurvey Amphitheater.
5. Unnamed deposit.

6. Unnamed deposit.
7. Little Basin.
8-9. Unnamed deposits.
10-11. Groundhog Creek.
12. Alaska-Juneau mill tailings.

doubtedly small. The gold was derived from nearby lodes that
were too small and too lean for profitable mining.

In the 1860's, W. P. Blake (1868, p. 10–12) noted gold in bars
on both sides of the international boundary when he ascended the
Stikine River with Russian officials. Commenting on the only
occurrence on the Alaskan side of the border (15, fig. 34), he
stated: "The gold is said to be quite fine and is in thin scales."

Near Lituya (28–35, fig. 34) and Yakutat (16–25, fig. 34) Bays, beach sands that contain local concentrations of gold and other heavy metals have been mined intermittently since before the purchase of Alaska in 1867. Production was small, probably totaling no more than 3,700 ounces. A little fine platinum was recovered with the gold from Lituya Bay. Black sands composed in part of magnetite and ilmenite have been investigated as possible sources of iron and titanium by Federal agencies (Rossman, 1957; Thomas and Berryhill, 1962) and private interests, but the results were discouraging; the richest samples indicated tenors of less than 20 pounds iron and a maximum of 43.7 pounds of titania (TiO_2) per cubic yard of beach material in place. The heavy minerals, concentrated from glacial deposits subject to marine erosion, probably came from as yet undiscovered sources in the largely unexplored, extremely rugged Fairweather Range.

Elsewhere in southeastern Alaska, placer occurrences of metallic minerals are very scarce. Near Chichagof on western Chichagof Island is the McKallick (1, fig. 34) placer, an eluvial deposit exposed by a small stream that cut through a muskeg in an area where many gold lodes have been mined. On Baranof Island at Goddard Hot Springs (2, fig. 34), heavy-mineral concentrates of samples collected by the Geological Survey contained as much as 7 percent allanite and traces of monazite and scheelite. In the area now included in Glacier Bay National Monument, there have been a few attempts to recover gold from outwash in front of Brady Glacier (3, fig. 34) and in the Dundas River drainage basin (4, 5, fig. 34).

Detrital garnets derived from garnetiferous schist are sufficiently concentrated along the west coast of Kanagunut Island (36, fig. 34) to be considered a beach placer. The deposit, however, has not been sufficiently investigated to determine its economic potential.

YUKON RIVER REGION

The Yukon River region (pl. 1, figs. 37–39, 41, 44, 47, 52) is the area drained by the Yukon River and its tributaries from the Alaska-Yukon boundary to the Bering Sea and by the Unalakleet and Manopiknak Rivers and intermediate streams flowing into Norton Sound and the Bering Sea. The region is divided into 26 districts described below in alphabetical order.

ANVIK DISTRICT

The Anvik district (pl. 1, fig. 37) comprises the western drainage of the Yukon River between Koyukuk and (and including) the Koserefsky River and the area drained by streams flowing into Norton Sound between Unalakleet and St. Michael Bay.

The district is characterized by north- to northeast-trending ridges with crests generally between 2,000 and 4,000 feet in altitude separated by structurally controlled valleys having long, straight segments. The ridges near Norton Sound are somewhat lower than in the rest of the district.

The area is largely underlain by Cretaceous and Tertiary(?) volcanic rocks and by mainly Cretaceous clastic marine and nonmarine units (some coal bearing) of the Koyukuk sedimentary basin (Cass, 1959a, b, e; Harrington, 1918; Patton and Hoare, 1968). An older volcanic unit (upper Paleozoic and Mesozoic) includes diabase intrusives and serpentinite as well as mafic volcanic and metavolcanic rocks. A large Mesozoic felsic pluton and several smaller similar bodies are in the Anvik River basin. The area southeast of St. Michael is underlain by Tertiary(?) and Quaternary basalt, some of which must be very young, for primary volcanic features are only slightly modified. The district has not been glaciated. It is mainly in a zone of thick to thin permafrost; a well at Anvik penetrated 31 feet of perennially frozen ground.

The one lode deposit discovered in the Anvik district consists of traces of cinnabar and stibnite found in 1962 by members of the Geological Survey in hydrothermally altered intrusive rhyolite on Wolf Creek Mountain about 40 miles northeast of Marshall (Berg and Cobb, 1967, p. 195, 198).

The only commercialy mined placer deposit in the Anvik district was near the head of the Stuyahok River on a small tributary called Flat Creek (1, fig. 37). Very little is known about this deposit other than that it was discovered in 1918 and was finally mined out in 1940, when the operator moved his equipment to the Ruby district. Concentrates contained small amounts of cinnabar. No data on the amount of gold recovered are available, as production was included with that of mines in the neighboring Marshall district. The combined production of both districts through 1961, the last year for which information is available, was probably between 115,000 and 120,000 fine ounces of gold, of which possibly as much as one-fifth should be credited to the Anvik district. Harrington (1918, p. 62–63) stated that prospectors had found gold on numerous bars of the Anvik River and that platinum was also said to be present, but there are no reliable reports that gold was ever actually mined. Schrader and Brooks (1900, p. 28) reported rumors that gold had been found on the Unalakleet River and in the headwaters of the Anvik River.

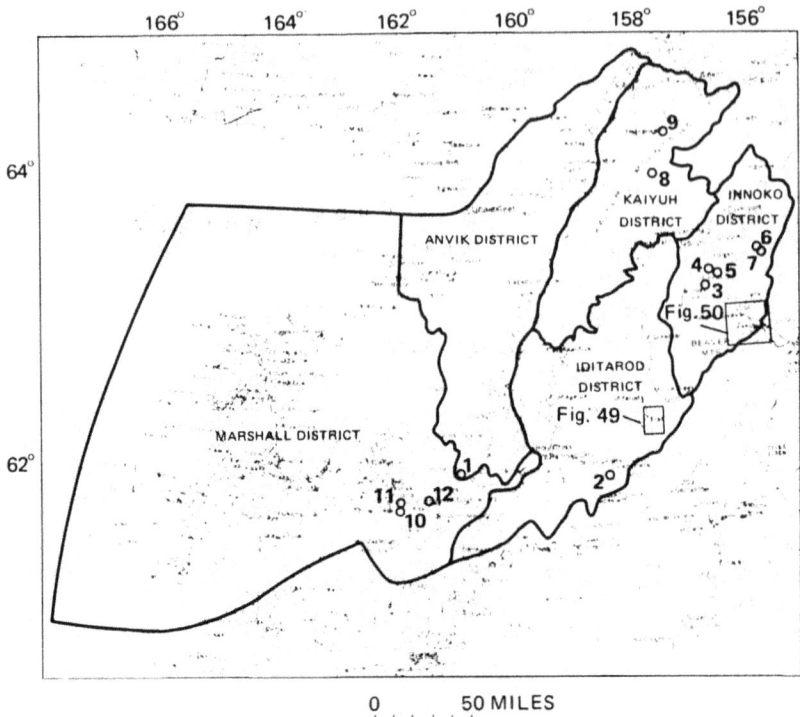

FIGURE 37.—Placer deposits in the Anvik, Iditarod, Innoko, Kaiyuh, and Marshall districts.

Anvik district

1. Stuyahok River (Flat Creek): Smith (1941b, p. 47), Joesting (1942, p. 27).

Iditarod district

2. Little Creek: Cady and others (1955, p. 120).

Innoko district

3. Boob Creek: Harrington (1919a, p. 349–350).

4. Iron Creek: Harrington (1919a, p. 350–351).

5. Esperanto Creek: Brooks and Capps (1924, p. 44). Keating Gulch: Smith (1939b, p. 49). Madison Creek: Harrington (1919a, p. 350–351).

6. Bear Creek: White and Killeen (1953, p. 16). Cripple Creek: Mertie (1936, p. 170–172), White and Killeen (1953, p. 16–18).

7. Colorado Creek: Mertie (1936, p. 172–173). Cripple Creek: Mertie (1936, p. 170–172), White and Killeen (1953, p. 16–18).

Fox Gulch: Eakin (1914a. p. 35), White and Killeen (1953, p. 16).

Kaiyuh district

8. Kluklaklatna (Little Mud) River: Mertie (1937a, p. 173).

9. Camp Creek (Tlatskokot): Alaska Department of Mines (1946, p. 38), Cass (1959b).

Marshall district

10. Willow Creek: Harrington (1918, p. 60–62), Hoare and Coonrad (1959b).

11. Disappointment Creek: Harrington (1918, p. 59). Elephant Creek: Harrington (1918, p. 59), Joesting (1942, p. 40), Hoare and Coonrad (1959b). Wilson Creek: Harrington (1918, p. 59), Hoare and Coonrad (1959b).

12. Bobtail Creek: Smith (1942b, p. 44), Joesting (1942, p. 27), Hoare and Coonrad (1959b). Buster Creek: Smith (1934b, p. 43). Montezuma Creek: Smith (1934b, p. 43), Hoare and Coonrad (1959b).

BLACK DISTRICT

The Black district (pl. 1, fig. 38) is bounded on the east by the
Alaska-Yukon boundary, on the south by the Yukon River, on the

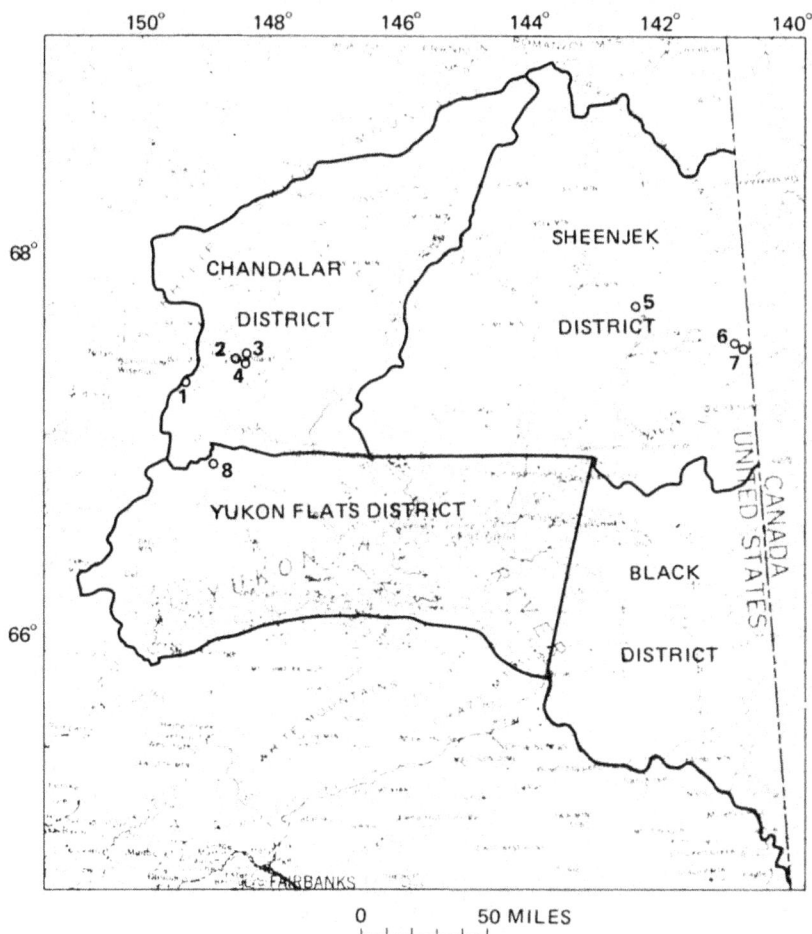

FIGURE 38.—Placer deposits in the Black, Chandalar, Sheenjek, and Yukon
Flats districts.

Black district

No placer occurrences.

Chandalar district

1. Dennys Gulch: Freeman (1963, p. 31),
   Brosgé and Reiser (1964).

2. Tobin Creek: White (1952b, p. 11), Nelson,
   West, and Matzko (1954, p. 17-18).

3. Big Squaw Creek: Mertie (1925, p. 259-
   260, 263), Nelson, West, and Matzko
   (1954, p. 16-19). Little Squaw Creek:
   Mertie (1925, p. 254-259, 263), Nelson,
   West, and Matzko (1954, p. 17-18).

Chandalar district—Continued

4. Big Creek: Mertie (1925, p. 260-261, 263).
   St. Marys Gulch: Maddren (1913, p. 116).

Sheenjek district

5. Pass Creek: W. P. Brosgé (oral commun.,
   Aug. 18, 1970).

6. Rapid River tributary: White (1952a, p. 8).

7. Sunaghun (Sunagun) Creek: White (1952a,
   p. 8-9).

Yukon Flats district

8. Slate Creek: Alaska Department of Mines
   (1956, p. 91). Trout Creek: Berg and
   Cobb (1967, p. 240).

west by an arbitrary line between Graphite Point on the Porcupine River and Circle on the Yukon River, and on the north by the divide between the Black and Porcupine Rivers.

The district consists generally of rounded ridges averaging 1,500–2,000 feet in altitude above which isolated summits project a few hundred feet and into which the principal rivers have cut wide, flat-floored valleys. A few peaks in the southeastern part of the district reach elevations of over 4,500 feet. The western boundary of the district roughly follows the eastern edge of the swampy Yukon Flats, which generally are less than 800 feet above sea level.

Unmetamorphosed sedimentary and volcanic rocks that range in age from Precambrian to Cretaceous underlie most of the Black district (Brabb, 1969; Brabb and Churkin, 1965, 1969). The youngest consolidated rock in the district is Tertiary or Quaternary basalt near a large lake (which may occupy an old caldera) 15 miles west of the mouth of the Salmon Fork of the Black River (Brabb, 1969). Quaternary fluvial deposits and loess floor the major valleys and blanket much of the northwest part of the district. The area was not glaciated but is in zones of discontinuous and thick to moderately thick permafrost.

The only metallic mineral resource in bedrock is hematite in the Precambrian Tindir Group (Kimball, 1969a). Williams (1951, p. 2) reported that "rich telluride float" was said to have been found in the early 1920's just north of the Yukon River about 28 miles upstream from Circle, but that searches for the source had been fruitless. In the same general area, he noted anomalous radioactivity during the course of a prospect examination but could not find any identifiable radioactive minerals.

No placer deposits have been found in the Black district.

BONNIFIELD DISTRICT

The Bonnifield district (pl. 1, fig. 39) is the area drained by southern tributaries of the Tanana River between and including the Teklanika and Little Delta Rivers.

The district includes part of the Alaska Range. Mount Deborah, Hess Mountain, and Mount Hayes are more than 11,900 feet in altitude, but most of the crest of the range is only about 5,000 feet above sea level. The mountains are flanked on the north by foothills that slope down to lowlands generally well below 1,000 feet in elevation.

Most of the district is underlain by Precambrian(?) and Paleozoic schists of both sedimentary and igneous origin, quartzite, and marble that were intruded by Mesozoic and Tertiary stocks

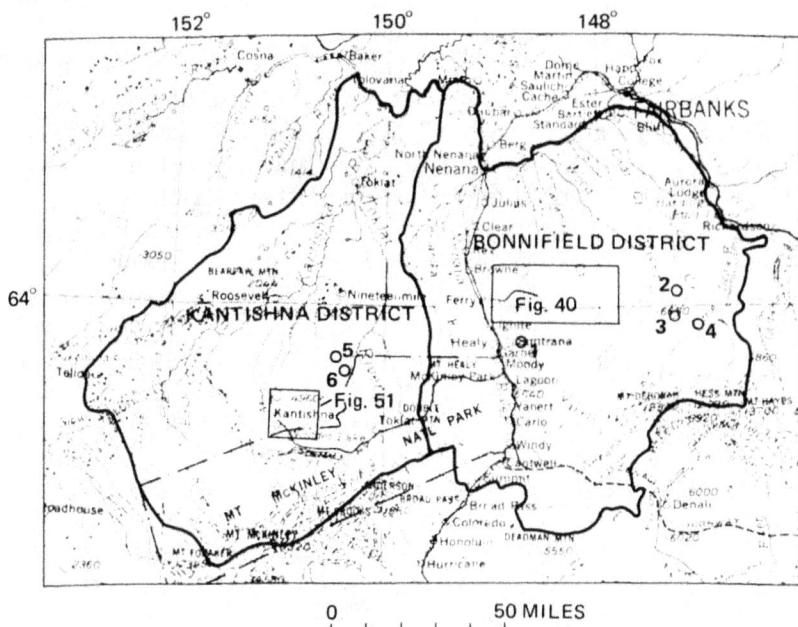

FIGURE 39.—Placer deposits in the Bonnifield and Kantishna districts.

Bonnifield district

1. Alaska Creek, French Gulch, and Gagnon (Home) Creek: Maddren (1918, p. 376, 378).
2. Caribou Creek: Capps (1912, p. 52).
3. Dry Creek: Capps (1912, p. 51). Newman Creek: Capps (1912, p. 52).

4. Portage Creek: Capps (1912, p. 52).

Kantishna district

5. Crooked Creek: Smith (1942b, p. 49).
6. Little Moose Creek: Capps (1919a, p. 3), Joesting (1942, p. 39). Stampede Creek: Joesting (1942, p. 39), White (1942, p. 335).

and dikes ranging in composition from felsic to mafic, granodiorite, dacite, andesite, and basalt (Péwé and others, 1966; Wahrhaftig, 1958, 1968, 1970a–1970h). The extreme southern part of the district is underlain by relatively unmetamorphosed Paleozoic and Mesozoic rocks similar to rocks in the northern part of the adjacent Cook Inlet-Susitna River region. Several thousand feet of Tertiary clastic and volcanic rocks crop out along the north flank of the Alaska Range. The youngest Tertiary formation commonly is poorly consolidated.

Deposits of several Pleistocene glaciations have been identified, particularly in the Nenana River valley. Except for several ice tongues that extended beyond the mountains, the lowlands were not glaciated, although thick glaciofluvial deposits and loess derived from them cover bedrock except in a few isolated groups of hills. The Mount Deborah-Mount Hayes massif is ice covered and is the source of several valley glaciers, two of which are more than 15 miles long. The district is in zones underlain by discontinuous or isolated masses of permafrost.

Lodes in the Bonnifield district are chiefly sulfide disseminations, veins, and lenses in schist near the borders of felsic intrusive bodies (Berg and Cobb, 1967, p. 198, 200–203, fig. 31). Small amounts of gold with subordinate silver and a few tons of antimony ore constitute the lode production of the district.

Placer gold was discovered on several creeks in the Bonnifield district in 1903 and 1904 by prospectors from Fairbanks, about 60 miles to the north. Bonnifield Creek (24, fig. 40), the stream for which the district was named, has been one of the least productive gold-bearing streams in the area. The district has been a small but steady producer; the total output through 1960 was probably 45,000 or 50,000 fine ounces. As published tables combined the production of this district with that of the adjoining Kantishna district for several years, a total figure for either is at best an approximation. The ultimate sources of the gold and associated heavy minerals were quartz veins and mineralized zones, similar to lodes that were mined or prospected, in pre-Mesozoic schist; much of the placer gold probably was reconcentrated from poorly consolidated Tertiary gravel, which nearly everywhere contains very small amounts of gold. Heavy minerals, in concentrates include, in addition to gold, various sulfides identified in nearby lodes, scheelite from Moose (1, fig. 40), Little Moose (2, fig. 40), Eva (3, fig. 40), Grubstake (16–18, fig. 40), and Gold King (21–23, fig. 40) Creeks, cassiterite from Moose Creek (1, fig. 40), cinnabar from Moose (1, fig. 40), California (4–5, fig. 40) and Grubstake Creeks, and platinum-group metals from Moose (1, fig. 40) and California Creeks. Most of the deposits are stream placers, although gravel on a few benches was mined. Mechanized equipment has been used on several creeks and a ditch about 6 miles long brought water to a hydraulic mine on Gold King Creek (22, fig. 40), but most of the miners used fairly primitive methods. In 1968, the only placer mining reported in the Bonnifield district was a small nonfloat operation on Platte Creek (9, fig. 40).

CHANDALAR DISTRICT

The Chandalar district (pl. 1, fig. 38) is the area drained by the Chandalar River and its tributaries above the village of Venetie. Its northern boundary is the crest of the Brooks Range.

The Brooks Range slopes southeastward from a crest line 5,000–7,000 feet in elevation to less rugged mountains in which few summits are above 5,000 feet. Broad valleys of some of the

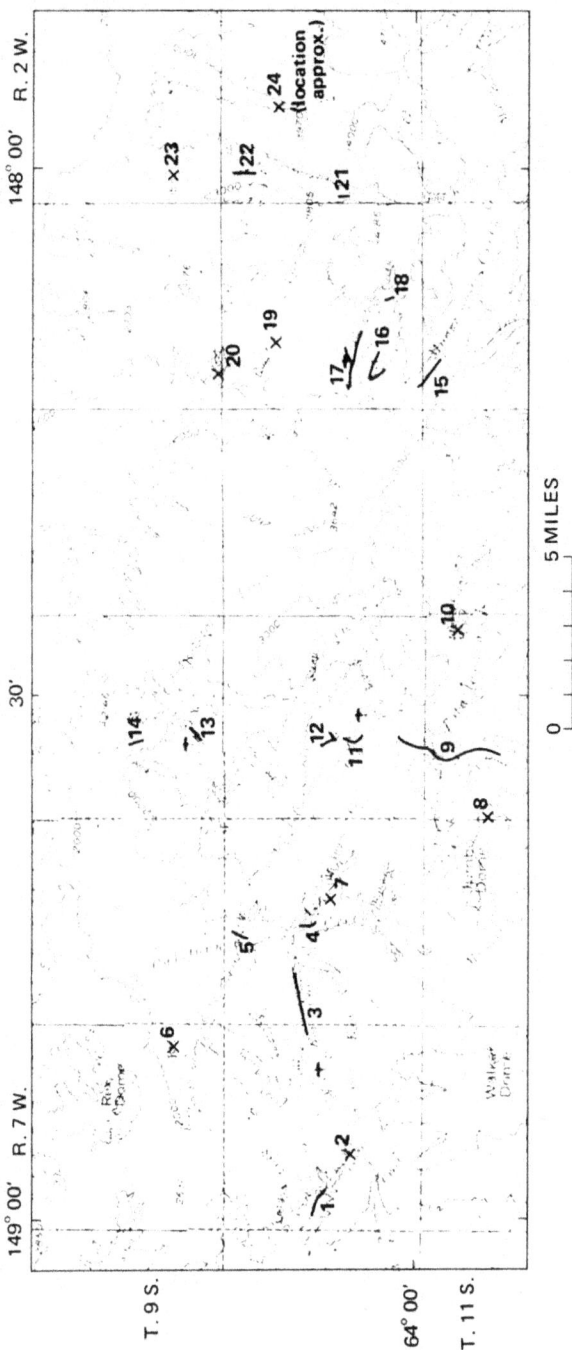

FIGURE 40.—Placer deposits in the Totatlanika River area.

1. Moose Creek: Overbeck (1918, p. 335), Maddren (1918, p. 365–368), Joesting (1942, p. 20, 27, 34, 39).

2. Little Moose Creek: Maddren (1918, p. 365–368), Joesting (1943, p. 20).

3. Eva Creek: Maddren (1918, p. 384–386), Joesting (1942, p. 39, 41).

4-5. California Creek: Capps (1912, p. 46), Maddren (1918, p. 380, 383), Joesting (1942, p. 20, 27).

6. Rex Creek: Maddren (1918, p. 380–383).

7. McAdam Creek: Moffit (1933, p. 345).

8. Marguerite Creek: Smith (1942b, p. 48).

9. Fox Gulch: Maddren (1918, p. 396–397), Homestake Creek (including Platte Creek): Capps (1912, p. 44–46), Maddren (1918, p. 395–397).

10. McCuen Gulch: Maddren (1918, p. 387–388, 397–398).

11. Totatlanika Creek (River): Maddren (1918, p. 388, 391–394).

12. Fourth of July Creek: Maddren (1918, p. 393–394). Totatlanika Creek (River): Maddren (1918, p. 388, 391–394).

13. Daniels Creek: Maddren (1918, p. 388–391).

14. Totatlanika Creek (River): Maddren (1918, p. 388, 391–394).

15. Moose Creek: Moffit (1933, p. 345).

16-18. Grubstake Creek: Capps (1912, p. 48), Maddren (1918, p. 399–400), Joesting (1942, p. 27, 39).

19. Roosevelt Creek: Maddren (1918, p. 399–400).

20. Hearst Creek: Maddren (1918, p. 899–400).

21-23. Gold King Creek: Maddren (1918, p. 400–401), Joesting (1942, p. 89).

24. Bonnifield Creek: Smith (1937, p. 46).

major forks of the Chandalar River merge downstream into the Yukon Flats, a swampy area of low relief that is generally less than 1,000 feet above sea level.

The district is underlain mainly by Paleozoic and Mesozoic sedimentary and mafic volcanic rocks. Basalt flows with an aggregate thickness of about 1,000 feet, exposed south of the Chandalar River across from the mouth of the East Fork, are of possible Tertiary age. Devonian low-grade metamorphic rocks crop out in a belt across the southern part of the district (Brosgé and Reiser, 1962, 1964, 1965). Mafic igneous rocks, probably at least in part of Jurassic age (Reiser and others, 1965), are in the eastern part of the district, and Jurassic and (or) Cretaceous granitic plutons, some with large contact aureoles, occur along the southern boundary of the district and near Chandalar Lake.

Except for some of the lower areas in the southern part, the district was covered by Pleistocene glaciers, remnants of which are still on some of the high peaks of the Brooks Range. Continuous permafrost underlies most of the district.

Lodes in the Chandalar district contain gold and various sulfides, including stibnite, galena, and sphalerite (Berg and Cobb, 1967, p. 203, fig. 30). The only production was an undetermined but probably small amount of gold, mainly from the Little Squaw mine near locality 3, figure 38. A neighboring property, the Mikado, is now (1970) being developed on a larger scale. These deposits consist of many steeply dipping auriferous quartz veins in schist cut by gneissic granite.

Placer gold was discovered in about 1906 in streams draining the area where lode deposits were soon to be discovered. Production figures are incomplete and ambiguous, but on the basis of published data, it seems reasonable that the total placer production of the district since 1906 has been about 25,000–30,000 fine ounces.

Most of the mining has been on Tobin, Big, Little Squaw, and Big Squaw Creeks and St. Marys Gulch, a tributary of Big Crook (2–4, fig. 38), all of which drain the area where lode deposits have been developed. Most mining was by individuals or small groups of men operating drift mines or using hand methods; mechanical equipment was used by one operator on Big Creek for several years in the late 1950's and possibly more recently. Placer concentrates from these streams contained a large suite of heavy minerals in addition to gold. The first recorded occurrence of monazite in Alaska was on Big Creek. Other heavy minerals include magnetite, hematite, rutile, pyrite, arsenopyrite, chalcopyrite, galena, stibnite, molybdenite, scheelite, and uranothorianite. Little Squaw Creek was dammed by ice that came down the North Fork of the Chandalar River and spilled over and filled the western part of a pass that separates the hills at the head of the creek from foothills of the Brooks Range. As a consequence of this damming, there are two generations of placers on Little Squaw Creek, one preglacial and the other postglacial. Big Squaw Creek probably had a similar Pleistocene history, but the valley of Big Creek, though occupied by local ice, was not greatly modified and has only one generation of placers.

Gold has also been mined somewhere on Dictator Creek (Smith, 1933a, p. 41), which flows into the Middle Fork of the Chandalar River about 13 miles south of the Little Squaw mine, and on Dennys Gulch (1, fig. 38), where one man worked for many years. Both of these streams flow in areas underlain by quartz-mica schist that at Dennys Gulch contains many thin discontinuous pyritiferous quartz veins.

CHISANA DISTRICT

The Chisana district (pl. 1, fig. 41) is the Alaskan part of the area drained by the upper White River and its tributaries and by the southern tributaries of the Tanana River above and including the Nabesna River.

From south to north it includes the summit areas and north slopes of the Wrangell Mountains, the Nutzotin Mountains and the southeastern part of the Mentasta Mountains (parts of the Alaska Range), and a lake-studded lowland above which rise several isolated patches of hills. Except for lowlands near the Tanana River, the district was covered by ice during at least two Pleistocene glaciations. The larger rivers all head in glaciers in the high mountains and carry heavy sediment loads. Most of the area is underlain by isolated masses of permafrost, some of which was encountered in placer workings.

The Chisana district is underlain mainly by thick sequences of Paleozoic and Mesozoic sedimentary and volcanic rocks intruded by several large Mesozoic dioritic plutons. There are also a few small intrusive bodies of probably Tertiary age. Several remnants of formerly more extensive Tertiary gravel locally overlie the older rocks. A thick sequence of Tertiary and Quarternary lava flows forms the high mountains in the southern part of the district. Unconsolidated glacial and fluvial deposits cover most of the lowlands.

Metallic lode deposits are scattered throughout the highland parts of the district (Berg and Cobb, 1967, p. 205, 208–209, fig. 32) except in areas underlain by lava. Between 1930 and World

War II, the Nabesna mine produced gold-silver-copper ore valued at $1,870,000. No production was recorded from any other lode deposit in the district. Of great current (1970) interest are two porphyry copper-molybdenum deposits that are being extensively investigated by private interests.

Placers in the Chisana district, almost all within a few square miles in the Bonanza Creek area (fig. 42), have been the source of about 45,000–50,000 ounces of gold since 1913 (Matson, 1969b, p. 2), 1914 and 1915 being the most productive single years. Bonanza and Little Eldorado Creeks (2, fig. 42) and Gold Run Creek and its tributaries (6, fig. 42) were the most extensively mined streams. According to Capps (1916a, p. 94–99), most of the gold in these creeks was reconcentrated from Tertiary gravel, a 200-foot-thick remnant of which is preserved as a cap on Gold Hill. The gold in the Tertiary gravel probably was derived from small veins in Paleozoic and Mesozoic bedded rocks, particularly in contact zones around plutons. All of this second-generation placer gold is worn and smooth. Bright rough gold from Big Eldorado Creek (4, fig. 42), some with crystal faces preserved and some with pieces of quartz attached, was derived from local lode sources. Gold has been found, but not mined commercially, in other streams in the Bonanza Creek area and in other parts of the district, but very little is known about most of these occurrences.

Native copper, a common constituent of concentrates from some streams in the Chisana district, was probably derived from amygdules in Paleozoic lava flows. In 1902, natives at Cross Creek (a few miles northwest of Chisana) showed copper nuggets to Mendenhall and Schrader (1903, p. 39–40). These nuggets were said to have come from a place called Tinast Gulch that is probably in the Cross Creek area; its exact location is not now known. Native copper has also been found in concentrates from Bonanza, Bryan (4, fig. 41), and Chathenda (1, fig. 42) Creeks. Other heavy minerals from Bonanza Creek, the only stream from which concentrates have been studied, include native silver, galena, and small amounts of cinnabar and molybdenite.

Placer mining in the Chisana district, first of creek gravels and later of bench and old channel deposits of Bonanza and Little Eldorado Creeks, has always been on a small scale with simple equipment. The remoteness of the area, shortages of water on some streams, and the small extent of the deposits all prevented the development of large operations. There has been little activity since World War II; the last reported mining was a two-man nonfloat operation in 1965.

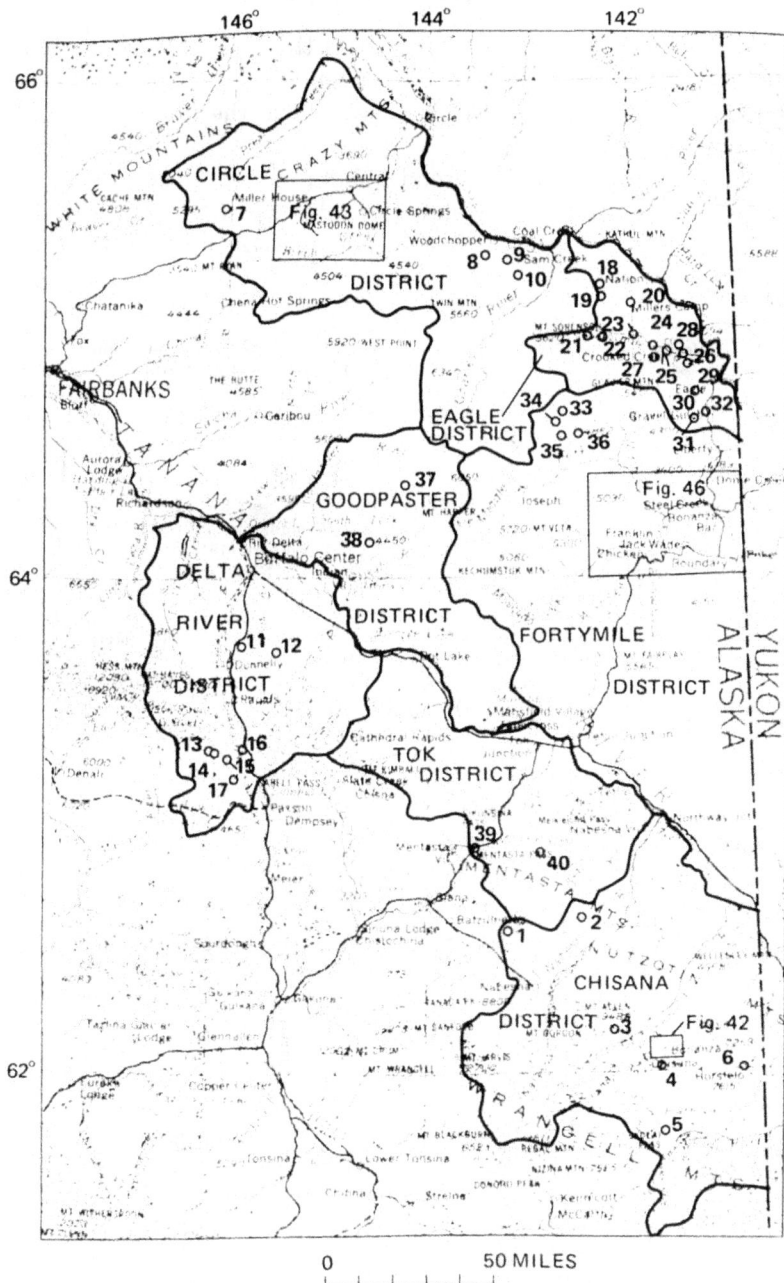

FIGURE 41.—Placer deposits in the Chisana, Circle, Delta River, Eagle, Fortymile, Goodpaster, and Tok districts.

Chisana district

Creek: Moffit (1954a, p. 201).
eek: Moffit (1941, p. 155).

Chisana district—Continued

2. Cheslina River (Creek): Moffit (1941, p. 153–154).

CIRCLE DISTRICT

The Circle district (pl. 1, fig. 41) is the area drained by north-flowing tributaries of the Yukon River from (and including) the Charley River to Circle, Birch Creek above the latitude of Circle, and most of Preacher Creek.

Most of the district is a broad upland of nearly accordant ridge crests 3,000–5,000 feet in elevation above which domes and isolated summits rise as much as 1,400 feet. A narrow trough ex-

---

FIGURE 41.—Continued.

Chisana district—Continued

3. Notch Creek: Martin (1919, p. 36).
4. Bryan Creek: Capps (1916a, p. 115–116).
5. Lime Creek: Capps (1916a, p. 116).
6. Horsfeld (Horsfall) Creek: Cairnes (1915, p. 132).

Circle district

7. Bachelor Creek: Prindle and Katz (1913, p. 149–150).
8. Alice Gulch: Brooks (1907c, p. 203), Cobb (1967e, p. 1). Iron Creek: Mertie (1938a, p. 255–256). Mineral Creek: Mertie (1938a, p. 255–257), Cobb (1967e, p. 1). Woodchopper Creek: Mertie (1938a, p. 254–256), Mertie (1942, p. 246–250, 257–259).
9. Boulder Creek: Mertie (1942, p. 251), Cobb (1967e, p. 1). Coal Creek: Mertie (1942, p. 246–251).
10. Sawyer Gulch: Mertie (1942, p. 250), Cobb (1967e, p. 1).

Delta River district

11. Ober Creek: Wedow, Killeen, and others (1954, p. 18).
12. McCumber Creek: Moffit (1942, p. 143–144). Morningstar Creek: Smith (1933a, p. 34).
13. Broxon Gulch: Rose (1965a, p. 35). Specimen Creek: Rose (1965a, p. 35).
14. Rainy Creek, West Fork: Rose (1965a, p. 34).
15. Rainy Creek: Rose (1965a, p. 2, 34).
16. Delta River: Wedow, Killeen, and others (1954, p. 18).
17. Delta River: Moffit (1912, p. 65).

Eagle district

18. Nugget and Surprise Creeks: Mertie (1938a, p. 204).
19. Dome and Eagle Creeks: Mertie (1938a, p. 204).
20. Fourth of July Creek: Mertie (1942, p. 246–250, 257–259). Ruby Creek: Prindle (1913b, p. 79).
21. Alder Creek: Prindle (1913b, p. 79), Mertie (1938a, p. 193–194).
22. Flume Creek: Prindle (1905, p. 57). Nugget Creek: Mertie (1938a, p. 193–194). Seventymile River: Mertie (1938a, p. 191–195).

Eagle district—Continued

23. Barney Creek: Mertie (1938a, p. 195). Pleasant (Placer) Creek: Ellsworth and Davenport (1913, p. 219–220).
24. Canyon Creek: Malone (1962, p. 50–51). Seventymile River: Mertie (1938a, p. 191–195). Sonickson Creek: Prindle (1905, p. 55–56). Washington (Little Washington) Creek: Ellsworth and Davenport (1913, p. 219).
25. Broken Neck Creek: Mertie (1938a, p. 195–196). Crooked Creek: Mertie (1938a, p. 196–198). Seventymile River: Mertie (1938a, p. 191–195), Mertie (1942, p. 257–259).
26. Seventymile River: Mertie (1938a, p. 191–195).
27. Mogul Creek: Joesting (1942, p. 27), unpub. data.
28. Fox Creek: Mertie (1938a, p. 198–199), Joesting (1942, p. 32). Lucky Gulch: Mertie (1938a, p. 198), Joesting (1942, p. 20).
29. Rock Creek: Ellsworth and Parker (1911, p. 171).
30. Wolf Creek: Smith (1941b, p. 54), unpub. data.
31. American Creek (including Discovery Fork): Mertie (1938a, p. 199–201). Star Gulch: Mertie (1938a, p. 201).
32. Dome Creek: Smith (1933a, p. 38–39).

Fortymile district

33. Ben Creek: Wedow, White, and others (1954, p. 20), Alaska Division of Mines and Minerals (1967, p. 83). Slate Creek: Wedow, White, and others, (1954, p. 20).
34. Ruby Creek: Wedow, White, and others (1954, p. 20).
35. Gold Run: Prindle (1913b, p. 80).
36. Fourth of July Creek: Wedow, White, and others (1954, p. 19).

Goodpaster district

37. Tibbs Creek: E. H. Cobb (this rept.).
38. Michigan Creek: Brooks (1918, p. 60).

Tok district

39. Mentasta Pass: Mendenhall and Schrader (1903, p. 47).
40. Moose Creek: Moffit (1954a, p. 190).

142° 00'　　　　　　　R. 19 E.　　　　　141° 50'　　　　R. 20 E.

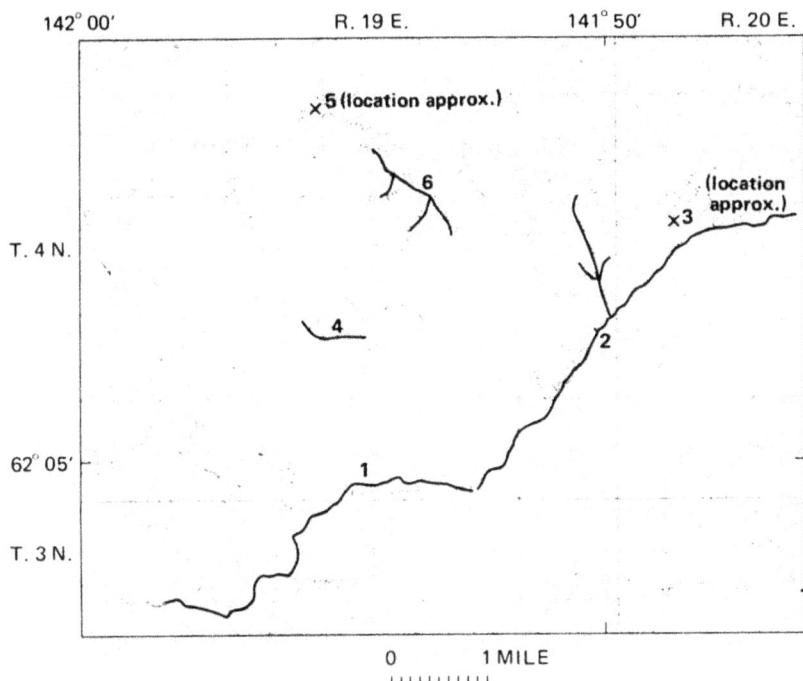

FIGURE 42.—Placer deposits in the Bonanza Creek area.

1. Chathenda (Johnson) Creek: Capps (1916a, p. 114–115). Dry Gulch: Capps (1916a, p. 115).
2. Bonanza Creek: Capps, 1916a, p. 99–109. Moffit (1943, p. 170–174). Little Eldorado Creek: Capps, (1916a, p. 95, 104–106, 109–110, 115), Moffit (1943, p. 173). Skookum Creek: Capps (1916a, p. 110–111). Snow Gulch: Capps (1916a, p. 115).
3. Coarse Money Creek: Moffit (1943, p. 173).
4. Big Eldorado Creek: Capps (1916a, p. 113–114).
5. Chavolda (Wilson) Creek: Capps (1916a, p. 116).
6. Discovery Pup: Moffit (1943, p. 173). Gold Run Creek: Capps (1916a, p. 111–112). Poorman Creek: Capps (1916a, p. 112, 116).

tends northwestward from the Alaska-Yukon boundary parallel to the Yukon River, from which it is separated by a line of low hills. The trough merges northwestward with the Yukon Flats, a low, lake-studded area less than 1,000 feet above sea level.

Only that part of the Circle district east of long 144° W. and north of lat 65° N. has been studied geologically (Brabb and Churkin, 1969) since the work of Mertie (1937b, 1938a, 1942) in the 1930's. The upland area is a complex metamorphic terrane of Precambrian(?) and Paleozoic gneiss, schist, and crystalline limestone that was invaded by mainly Mesozoic felsic, mafic, and ultramafic intrusive bodies, the largest of which is the Charley River batholith, which underlies much of the southeastern part of the district (Brabb and Churkin, 1969; Mertie, 1937b, 1938a).

Tertiary continental rocks underlie the trough southwest of the Yukon River (Mertie, 1942). Quaternary deposits are mainly alluvium and terrace gravels along streams and widespread loess. Glacial deposits occur near some of the highest mountains, the only parts of the area that were covered by Pleistocene ice. The highland parts of the district are characterized by discontinuous permafrost, and the lowlands, by thick or thin to moderately thick, generally continuous permafrost.

A few prospects in the Circle district have been explored for copper, gold, or lead, but none was productive (Berg and Cobb, 1967, p. 210, fig. 32). Possible sources of minerals reported from placer concentrates are auriferous quartz veins in metamorphic rocks on Porcupine Dome (about 3 miles southwest of locality 1, fig. 43), mineralized fault zones, a small wolframite-bearing vein in schist on Deadwood Creek (7, fig. 43), and occurrences of lead, zinc, and copper minerals and scheelite in granitic rocks near Miller House and Circle Hot Springs.

A report of the discovery of gold somewhere on Birch Creek in 1893 was followed by a rush of prospectors the next year, during which promising prospects were found on most of the tributaries of Crooked Creek. By 1896, all of the principal streams of the Miller House-Circle Hot Springs area (fig. 43) from which large amounts of gold were later produced had been staked and were being mined. A few years later gold was discovered on tributaries of the Yukon River about 40 miles to the east. Mining in the Circle district has been reported for every year since 1894. Total production through 1961 was about 735,000 fine ounces of gold, about 3.6 percent of the total for the State. Data are not available for the years since 1961, but production was probably sufficient to raise the total to close to 750,000 fine ounces. In 1964, the last large operation in the district, a dredge on Woodchopper Creek (8, fig. 41), closed down. By 1968, mining activity had dwindled to a two-man nonfloat operation on Deadwood Creek.

The heavy metals in the placers were derived from mineralized veins in Precambrian(?) and Paleozoic metamorphic rocks and younger felsic plutons and possibly from contact-metamorphic deposits in border zones. Gold in the deposits on Coal (9, fig. 41) and Woodchopper (8, fig. 41) Creeks and neighboring streams may have been reconcentrated from Tertiary conglomerates formed largely by the erosion of nearby metamorphic rocks. Both stream gravels and low bench deposits have been mined, and practically all placer-mining methods have been employed. Dredges were used on most of the major creeks. The first one was moved

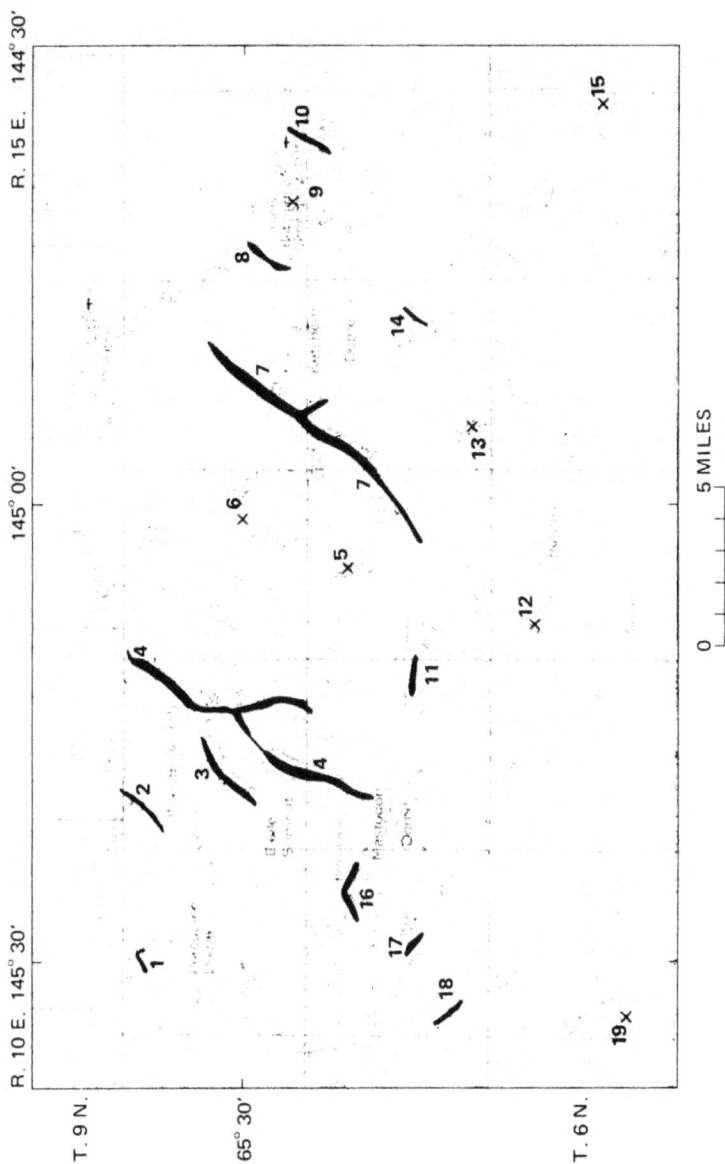

FIGURE 43.—Placer deposits in the Miller House-Circle Hot Springs area.

FIGURE 43.—Continued.

1. Porcupine Creek: Mertie (1938a, p. 224-226). Yankee Creek: Mertie (1938a, p. 225).

2. Bonanza Creek: Mertie (1932, p. 169), Mertie (1938a, p. 225-228).

3. Miller Creek: Mertie (1938a, p. 222-224).

4. Crooked Creek: Nelson, West, and Matzko (1954, p. 11). Independence Creek: Mertie (1938a, p. 210-211, 218-222), Nelson, West, and Matzko (1954, p. 13-14). Mammoth Creek: Mertie (1938a, p. 205-208). Mastodon Creek: Mertie (1938a, p. 205-208), Joesting (1942, p. 32).

5. Greenhorn Creek (Gulch): Mertie (1938a, p. 250).

6. Boulder Creek: Mertie (1932, p. 164).

7. Deadwood Creek: Johnson (1910), Brooks (1918, p. 56), Mertie (1938a, p. 235-245), Joesting (1942, p. 27, 37, 41), Nelson, West, and Matzko (1954, p. 14). Switch Creek: Johnson (1910, p. 248-250), Mertie (1938a, p. 236-238, 245-248) Joesting (1942, p. 37, 41).

8. Holdem Creek: Smith (1936, p. 39). Ketchem Creek: Mertie (1938a, p. 248-250), Wedow, White, and others (1954, p. 5).

9. Hot Springs Creek: Wedow, White, and others (1954, p. 5).

10. Portage Creek: Nelson, West, and Matzko (1954, p. 11-15), Wedow, White, and others (1954, p. 4-5).

11. Harrison Creek, North Fork: Mertie (1938a, p. 231-235), Joesting (1942, p. 32).

12. Harrison Creek (including South Fork): Mertie (1938a, p. 232-233).

13. Squaw Creek (Gulch): Brooks (1907c, p. 195).

14. Bottom Dollar Creek: Mertie (1938a, p. 231-235). Half Dollar Creek: Smith (1941b, p. 41), Joesting (1943, p. 19-20). Two Bit Gulch: Ellsworth and Parker (1911, p. 164).

15. Birch Creek: Ellsworth and Davenport (1913, p. 213).

16. Eagle Creek: Mertie (1938a, p. 228-231).

17. Gold Dust Creek: Mertie (1938a, p. 231).

18. Butte Creek: Mertie (1938a, p. 231).

19. Frying Pan Creek: Ellsworth and Parker (1911, p. 160, 164).

from the Klondike district of the Yukon Territory to Mastodon Creek (4, fig. 43) in 1912. It was operated for only 2 years, as it was too small to dig to bedrock. Large dredges and nonfloat operations using draglines probably accounted for most of the production; there were also some large hydraulic mines. As most of the auriferous gravel was permanently frozen, it had to be thawed in advance of mining, generally by groundsluicing off the overburden.

Heavy minerals, in addition to gold, in placer concentrates from the Miller House-Circle Hot Springs area include magnetite, ilmenite, garnet, and zircon; pyrite, pyrrhotite, and other sulfide minerals; and cassiterite, scheelite, and wolframite. Cassiterite is widespread, having been reported from Porcupine Creek (1, fig. 43), Mastodon and Independence Creeks (4, fig. 43), Deadwood Creek, and small streams, including Ketchem (8, fig. 43), Portage (10, fig. 43), and Half Dollar (14, fig. 43) Creeks and the North Fork of Harrison Creek (11, fig. 43). It was saved only from operations on Deadwood Creek. Wolframite and, less commonly, scheelite were reported from several of the same creeks and from Switch (7, fig. 43) and Hot Springs (9, fig. 43) Creeks. A little wolframite recovered as a byproduct of gold mining on Deadwood Creek was sold when tungsten prices were high during World War I. Investigations of potential sources for radioactive materials in 1949 and 1952 (Wedow, White, and others, 1954; Nelson and others, 1954) resulted in the discovery of small amounts of uranium and (or) rare-earth minerals in concentrates from several streams, including Independence, Ketchem, Hot Springs, and Portage Creeks. A few grains of cinnabar, for which no lode source has been found, were reported from Deadwood Creek. Gold from Woodchopper Creek, like that from Fourth of July Creek and the Seventymile River in the Eagle district, contains alloyed platinum, none of which was sold separately.

There has been considerable prospecting and some small-scale mining on streams other than those mentioned above or shown in figure 43, though the total production probably was small (Mertie, 1938a, p. 252; Mertie, 1942, p. 250). In the eastern part of the district, tributaries of Woodchopper Creek, in particular Mineral Creek and Alice Gulch (8, fig. 41), supported a thriving mining industry from 1898 to possibly as recently as 1925; most of the work was done before World War I. The valleys of Thanksgiving and Webber Creeks (west of Woodchopper Creek), upper Coal Creek, and Sam Creek and the Charley River (east of Coal Creek) were explored and some gold was produced. The sites of mining were not described well enough to plot them on the map of the area (fig. 41). Prospecting in the Preacher Creek basin in the northwestern part of the district was generally unsuccessful, but some gold was found. The only place with recorded production was Bachelor Creek (7, fig. 41), where gravel on a low bench east of the stream was sluiced in 1910.

DELTA RIVER DISTRICT

The Delta River district (pl. 1, fig. 41) is the area drained by northward-flowing tributaries of the Tanana River between and including the Johnson River on the east and Delta Creek on the west.

The district is mainly in the Alaska Range and is marked by many peaks 8,000 feet or more in altitude. The drainage divide between the Delta and Gulkana Rivers is about 35 miles south of the higest peaks. From the crest of the Alaska Range, the land slopes northward to flats along the Tanana River where elevations are less than 1,500 feet.

The district is characterized by two geologic domains separted by the Denali fault, one of the major strike-slip tectonic elements of Alaska (Grantz, 1966; Moffit, 1954a; Rose, 1965a; Wahrhaftig, 1968). North of the fault the area is underlain mainly by Precambrian and Paleozoic metamorphosed sedimentary rocks and by Mesozoic dikes, sills, and monzonitic plutons. South of the fault a few exposures of metamorphic rocks have been reported, but most of the area is underlain by relatively unmetamorphosed Paleozoic and Mesozoic volcanic and sedimentary rocks and felsic to ultramafic dikes, sills, and stocks. Tertiary nonmarine rocks cover large areas north of the mountain front and smaller areas in the Alaska Range. The mountains were the site of several eposides of Pleistocene glaciation during which large tongues of ice advanced down the valley of the Delta River nearly to its confluence with the Tanana. Much of the high country is still ice covered, and large valley glaciers extend down to elevations well below 3,000 feet. Glacial and glaciofluvial deposits cover most of the district where slopes are not too steep, and loess derived from periglacial flood plains blankets much of the Tanana Valley. Most of the district is in a zone of discontinuous permafrost; some areas in the lowlands are underlain by moderately thick to thin permafrost.

Except for a little ore that was mined, but not shipped, from a molybdenite deposit about 15 miles north of Mount Hayes, there has been no lode mining in the Delta River district, though numerous occurrences of sulfide minerals have been found; most are associated with ultramafic or volcanic rocks south of the Denali fault (Berg and Cobb, 1967, p. 211–212, fig. 32). Some of these deposits and a quartz vein in schist a few miles north of the Denali fault carry a little gold, but not enough to have been mined.

No valuable placers have been found in the Delta River district. The output of gold has been so small that production figures were combined with those of other districts; total production probably was not much more than 1,000 ounces. The proximate source of gold in McCumber and Morningstar Creeks (12, fig. 41) and Ober Creek (11, fig. 41), and showings on Jarvis Creek (Wedow, Killeen, and others, 1954, p. 18), was probably gravel of Tertiary age. Monazite was identified in a concentrate sample from Ober Creek. At Rainy Creek (14, fig. 41), where the gold may have been reconcentrated from glacial deposits, placers were worked sporadically from 1900 until at least as recently as 1930. Rose (1965a, p. 35) found a tractor and other machinery and the remains of ditches on a fork of Broxon Gulch (13, fig. 41) and other signs of mining on a nearby creek. The results of the operations were not reported. In 1946, a company was said to be minig on the east side of the Delta River (16, fig. 41), but there was no sign of activity 2 years later. No other mining in the Delta River district has been reported since World War II.

EAGLE DISTRICT

The Eagle district (pl. 1, fig. 41) is the area drained by the southern tributaries of the Yukon River between (but excluding) the Charley River on the west and the Alaska-Yukon boundary on the east. The district is largely in the uplands between the Yukon and Tanana Rivers, an area of even-topped ridges 3,000–5,000 feet in elevation separated by generally narrow valleys and surmounted by isolated mountains that rise as much as 1,000 feet above the general level. A trough southwest of, and parallel to, the Yukon River is separated from the river by discontinuous low hills.

The upland areas are part of an extensive, mainly Precambrian(?) and Paleozoic metamorphic terrane that was invaded by Mesozoic and, to a lesser extent, Paleozoic and Tertiary plutons, dikes, and sills of felsic, mafic, and ultramafic composition (Brabb and Churkin, 1965, 1969; Clark and Foster, 1969a, Foster and Keith, 1968, 1969). The trough near the Yukon River is underlain by Tertiary and Upper Cretaceous continental conglomerate, finer grained clastic rocks, and thin coal beds. Between the trough and the Yukon River are unmetamorphosed sedimentary and volcanic rocks of Precambrian, Paleozoic, and Mesozoic age. Quaternary deposits in the district are mainly alluvium and terrace deposits along streams. Except for cirque glaciers on some of the higher mountains, the district was ice free during the Pleistocene Epoch. The area is mainly in a zone of discontinuous permafrost.

No workable lodes have been found in the Eagle district, although sulfide-bearing material was reported from Eagle Bluff near the mouth of Mission Creek (Berg and Cobb, 1967, p. 213). Anomalous stream-sediment samples and specimens of copper-stained rock were collected in 1968 by Clark and Foster (1969b) during a geochemical reconnaissance of part of the Seventymile River area.

Placer-gold deposits were discovered in the basins of American Creek and the Seventymile River in about 1895 and on Fourth of July Creek (20, fig. 41) in 1898. Within a few years, most of the creeks in which there since has been mining were staked. The Eagle district has been a consistent, though small, producer since about 1900. Total production of placer gold through 1960, the last year for which reliable records are available, was about 45,000 ounces. As the reported activity since 1960 was restricted in most years to a small operation on Alder Creek (21, fig. 41) (H. L. Foster, oral commun., Dec. 10, 1970), it is doubtful that the total would be significantly larger if brought up to date. All mining in the district has been on a relatively small scale, mainly using hand and hydraulic methods, although some mechanical equipment has been employed.

The ultimate source of most of the gold was mineralized quartz veins in metamorphic rocks. Another source was mineralized material associated with altered ultramafic rocks in a fault zone near Flume (22, fig. 41) and Alder (21, fig. 41) Creeks (Clark and Foster, 1969b, p. 3). Much of the gold in the placers came directly from veins, but some may have been immediately derived from Tertiary and possible Upper Cretaceous conglomerate and associated clastic rocks that had been in turn largely derived from local sources in the metamorphic terrane during earlier erosion cycles. As no local source has been found for the numerous chert pebbles in the conglomerates, however, some of the gold may be exotic. Platinum, derived from ultramafic rocks, occurs as discrete grains in concentrates from Lucky Gulch (28, fig. 41) and is alloyed with gold at Fourth of July Creek (20, fig. 41) and the Seventymile River (25, fig. 41) at the mouth of Broken Neck Creek. The alloyed platinum was not sold separately from the gold, but was claimed by the United States Treasury as seigniorage when the gold was processed. Cinnabar, though never recovered commercially, was a constituent of concentrates from Canyon Creek (24, fig. 41) and was found nearby in the drainage basin of Mogul Creek (27, fig. 41). Despite extensive searches, however, the lode source of the cinnabar could not be determined. Cassiterite occurs sparsely in concentrates from Fox Creek (28, fig. 41) and chromite in the placers of Wolf Creek (30, fig. 41). Native silver nuggets were recovered with gold from Crooked Creek (25, fig. 41), the site of one of the largest hydraulic mining operations in the area. In addition to the creeks mentioned, there was mining at the other localities in the Eagle district shown in figure 41.

FAIRBANKS DISTRICT

The Fairbanks district (pl. 1, fig. 44) is the area drained by the Chatanika River and northern tributaries of the Tanana River from Minto to and including Shaw Creek.

The district is a dissected plateau, 2,000–4,500 feet in altitude, that rises gently from west to east and is characterized by many wide valleys separated by broad rolling divides surmounted by rounded domes and a few mountainous areas that rise several hundred feet higher.

The following summary of the geology of the Fairbanks district is based on reports by Capps (1940), Mertie (1937b), Péwé (1955), and Prindle (1913a) and on those cited in the captions for figures 44 and 45.

The oldest rocks in the Fairbanks district are schist, crystalline limestone, quartzite, amphibolite, and gneiss of Precambrian and early Paleozoic age intruded by mainly Mesozoic plutons and dikes, most of which are granodiorite, quartz diorite, or porphyritic granite and quartz monzonite. Nearly all the domes and mountains that rise above the general upland surface are underlain by such rocks. Many serpentinized ultramafic bodies are in the upper valley of the Salcha River in the eastern part of the district. Tertiary basalt crops out on a low hill near Fairbanks. Except for a few local cirque glaciers on the highest mountains, the district was not ice covered, but the Quaternary history of the area is complex. The uplands are generally covered by a blanket of loess derived (and still being derived) from the proglacial flood plains of streams issuing from the Alaska Range many miles south of the Tanana River. Transported loess mixed with locally derived clastic material and vegetation chokes valleys and forms the frozen muck that overlies most of the placer-gold deposits. Frozen and skeletal remains of Pleistocene mammals are common in this material.

Most of the lodes in the Fairbanks district are concentrated in an area within about 25 miles of Fairbanks; a few have been

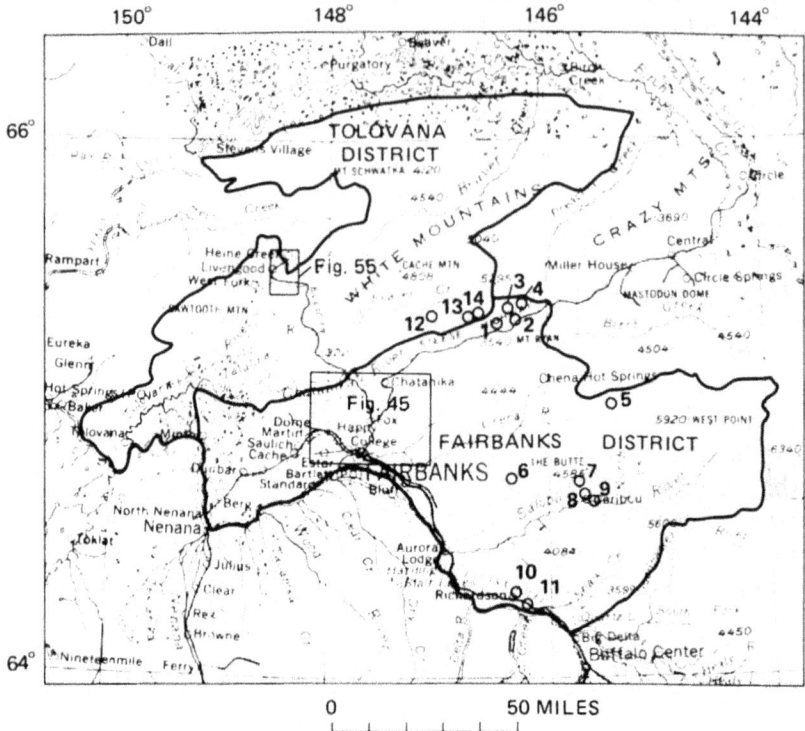

FIGURE 44.—Placer deposits in the Fairbanks and Tolovana districts.

Fairbanks district

1. Sourdough Creek: Joesting (1942, p. 14, 32), Nelson, West, and Matzko (1954, p. 11), Wedow, Killeen, and others (1954, p. 8).
2. Deep Creek: Wedow, Killeen, and others (1954, p. 8). Faith Creek: Smith (1942b, p. 39).
3. Hope Creek: Brooks (1907a, p. 37).
4. Charity Creek: Burand (1965, p. 3). Homestake Creek: Prindle (1910a, p. 209), Ellsworth (1912, p. 242).
5. Chena River (Van Curler's Bar): Ellsworth (1912, p. 244), Smith (1942b, p. 40). Palmer Creek: Joesting (1942, p. 39, 41). Shamrock Creek: Smith (1941b, p. 43).
6. Beaver Creek: Ellsworth and Davenport (1913, p. 208), Joesting (1942, p. 34). Pine (Pyne) Creek: Joesting (1942, p. 34).
7. Butte Creek: Prindle (1906, p. 123–125).

Fairbanks district—Continued

8. Caribou Creek: Prindle (1913b, p. 80), Joesting (1942, p. 34, 39).
9. No Grub Creek: Smith (1942b, p. 39).
10. Banner Creek: Saunders (1965, p. 2). Buckeye Creek: Wedow, Killeen, and others (1954, p. 11, 13), Saunders (1965, p. 4). Democrat Creek: Saunders (1965, p. 4). Hinkley Gulch: Saunders (1965, p. 2, 4). Moore Creek: Smith (1930a, p. 26).
11. Tenderfoot Creek: Prindle and Katz (1913, p. 141), Saunders (1965, p. 2, 4).

Tolovana district

12. Nome Creek: Ellsworth and Parker (1911, p. 165), Ellsworth (1912, p. 243–244). Ophir Creek: Ellsworth and Parker (1911, p. 165), Martin (1920, p. 38).
13-14. Nome Creek: Ellsworth (1912, p. 243–244), Smith (1942b, p. 38–39, 67), Wedow, Killeen, and others (1954, p. 8).

found elsewhere but none were productive (Berg and Cobb, 1967, p. 213–221, fig. 33). Total lode production to 1960 was 239,247 ounces of gold, 39,078 ounces of silver, 2,500–3,000 tons of antimony ore, and scheelite ore and concentrates containing several thousand units of WO_3. In discussing the gold lodes near Fairbanks, Chapman and Foster (1969, p. D1) stated: "The deposits are mainly concentrated in two areas within the district—the Pedro Dome-Cleary Creek area and the Ester Dome area. In the Pedro Dome-Cleary Creek area, they are localized along the Cleary Creek anticline. The tungsten deposits (predominantly scheelite) are principally in the Gilmore Dome-Tungsten Hill area; a small group of antimony deposits (stibnite) is in the area at the head of Vault and Treasure Creeks. Lode deposits are conspicuously absent outside these four areas." Lodes in other parts of the district, although less thoroughly studied, appear to be in similar geologic settings.

The Fairbanks district has been the leading producer of placer gold in Alaska, exceeding the total production of the Seward Peninsula region by more than a million ounces. Through 1961 the output was about 7,550,000 fine ounces of gold, or 37.2 percent of the total recorded Alaskan placer-gold production up to that time. Data for years since 1961 are not available, but as large dredges operated until the close of the 1963 season, the 1961 total was probably increased by about 100,000 ounces. Production credited to the Fairbanks district includes an undetermined, but probably relatively small, amount of the gold produced at Nome Creek, about 45 miles from Fairbanks in the Tolovana district.

Felix Pedro discovered valuable gold placers on the stream that now bears his name (14, fig. 45) in 1902, a year after Fairbanks had been established as a trading station on a slough of the Tanana River. Other discoveries soon followed, but development was slow, as most of the auriferous gravels were deeply buried by barren gravel and muck and the use of hoisting machinery and the development of transportation facilities first were necessary. Because water with a sufficient natural head for mining was scarce, pumps had to be employed. Once these initial difficulties were overcome, the area near Fairbanks boomed and many of the high-grade deposits were virtually mined out by elaborate drift mining and scraper operations. By the end of 1909, $9 million in gold (at $20.67 per ounce) had been recovered in each of 3 years; annual outputs never approached this in later years. The principal creeks, listed in order of productivity, were Cleary (31, fig. 45) and its tributaries, Ester (2, fig. 45), Fairbanks (32, fig. 45) and its tributaries, Dome (25, fig. 45), and Goldstream (14, fig. 45) and its tributaries. Each of these produced gold worth more than $4 million (at $20.67 per ounce) and several other streams accounted for at least $1 million each. The average tenor of the gravel mined in 1908 was about $5.60 per cubic yard (Prindle and Katz, 1913, p. 95), an indication of the richness of the placers worked in the early days of the development of the district.

After its culmination in 1909, annual production fell until 1928, when the first large dredges were installed following the consolidation of many separate holdings and the construction of a ditch that brought water from near the head of the Chatanika River. Cold water was used to thaw frozen ground ahead of dredges rather than the more elaborate boiler and steam-point systems that had been used for drift mining. From 1928 until the last dredges shut down at the end of the 1963 season, the Fairbanks area was dominated by large dredges working lean ground and reworking areas that had been mined by other methods. In addition, smaller dredges were used for many years, particularly on Fairbanks Creek and at Van Curler's Bar on the Chena River (5, fig. 44). Mining by methods other than dredging was carried on intermittently in several parts of the district outside the immediate Fairbanks area. The main centers were Sourdough Creek (1, fig. 44) and neighboring Faith, Hope, and Charity Creeks (2–4, fig. 44) and their tributaries; Caribou (8, fig. 44) and neighboring creeks in the Salcha River basin; and a group of small creeks in what was formerly called the Richardson or Tenderfoot district (10, 11, fig. 44). The total production from these outlying areas and the Chena River probably accounted for 100,000 to 200,000 ounces of gold. After 1963, placer mining in the district practically ceased; in 1968, only five operations employing a total of 13 men were reported to be active.

Most of the heavy minerals in the placers of the Fairbanks district were probably derived locally from lode deposits, as nearly all the valuable placers were found in areas where lodes containing most of the same minerals have been mined or prospected. A few minerals for which lode sources have not been found have been reported from concentrate samples. Cassiterite, in particular, is a minor constituent of concentrates from creeks in all parts of the district; scheelite is fairly common in several creeks in areas with no known lode tungsten resources. Both of these minerals have also been reported from other parts of the Yukon-Tanana area where lode sources have not been found.

1. Emma Creek: Ellsworth and Parker (1911, p. 158).

2. Ester Creek: Brooks (1907a, p. 30), Prindle (1908a, p. 29, 44–46), Prindle and Katz (1913, p. 103–105), Byers (1957, p. 188). Eva Creek: Brooks (1916a, p. 59). Ready Bullion Creek: Prindle (1908a, p. 45).

3–4. Cripple Creek: Prindle (1908a, p. 29, 44–45), Smith (1942b, p. 38–40), Joesting (1942, p. 32), Chapman and Foster (1969, pl. 1).

5. St. Patrick Creek: Eakin (1915b, p. 235).

6. Happy Creek: Smith (1942b, p. 39).

7. Little Nugget and Sheep Creeks: Chapman and Foster (1969, pl. 1).

8. Nugget Creek: Chapman and Foster (1969, pl. 1).

9. O'Connor Creek: Prindle (1908a, p. 39, 41), Prindle and Katz (1913, p. 106).

10–11. Big Eldorado Creek: Prindle and Katz (1913, p. 106).

12–13. Fox Creek: Prindle (1908a, p. 39, 41), Prindle and Katz (1913, p. 105–106), Byers (1957, p. 188, 210).

14. Engineer Creek: Prindle and Katz (1913, p. 105–106). First Chance Creek: Smith (1942b, p. 39), Joesting (1942, p. 32, 39). Flume Creek: Brooks (1916a, p. 59). Gilmore Creek: Prindle (1908a, p. 39–40), Hill (1933, p. 71), Joesting (1942, p. 39–40). Goldstream Creek: Prindle and Katz (1913, p. 105), Joesting (1942, p. 32, 39), Chapman and Foster (1969, pl. 1). Hill Creek: Brooks (1925. p. 19). New Years Pup: Chapin (1914a, p. 359). Pedro Creek: Prindle (1906, p. 111, 118), Joesting (1942, p. 32). Rose Creek: Prindle and Katz (1913, p. 113). Byers

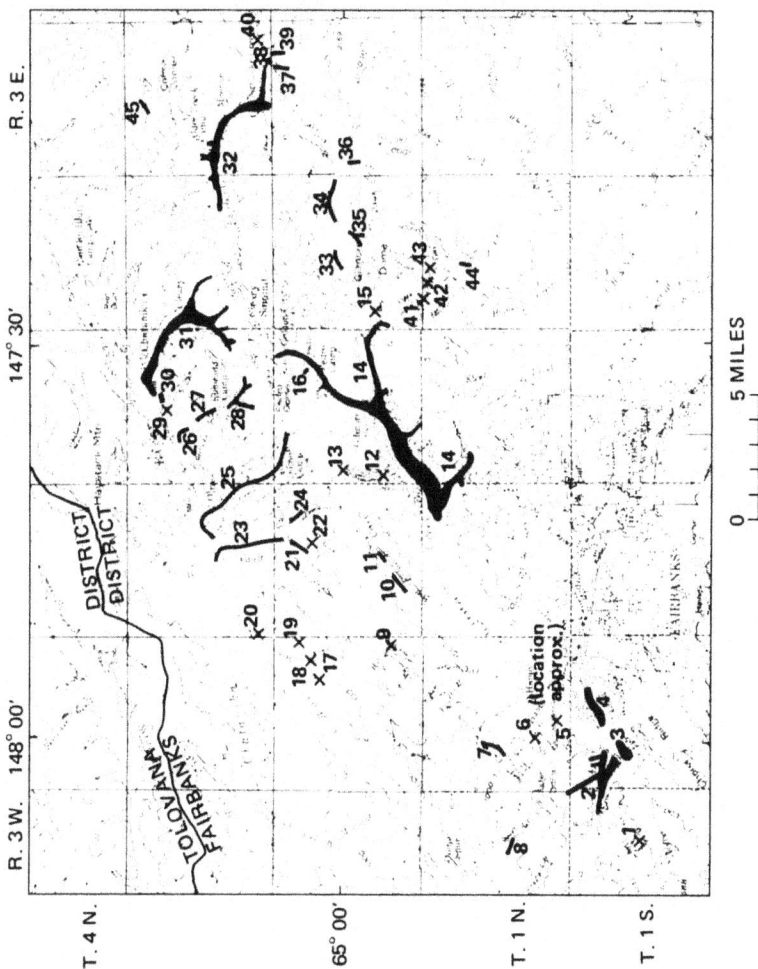

FIGURE 45.—Placer deposits in the Fairbanks area.

FIGURE 45.—Continued.

(1957, p. 188, 211). Steamboat Creek (Pup): Brooks (1916a, p. 59). Twin Creek: Prindle and Katz (1909, p. 188, 192).

15. Gilmore Creek: Prindle (1908a, p. 39–40), Hill (1933, p. 71), Joesting (1942, p. 39–40).

16. Steamboat Creek (Pup): Brooks (1916a, p. 59).

17-18. Our Creek: Prindle and Katz (1913, p. 101–102).

19. Unnamed deposit: Chapman and Foster (1969, pl. 1).

20. Sargent Creek: Chapman and Foster (1969, pl. 1).

21. Treasure Creek: Prindle and Katz (1913, p. 101), Chapman and Foster (1969, pl. 1).

22. Wildcat Creek: Prindle and Katz (1913, p. 101), Brooks (1916a, p. 59).

23-24. Vault Creek: Prindle (1908a, p. 29, 43–44), Prindle and Katz (1913, p. 101), Chapman and Foster (1969, pl. 1).

25. Dome Creek: Prindle and Katz (1913, p. 100–101), Joesting (1942, p. 32, 37), Joesting (1943, p. 20, 28), Byers (1957, p. 188, 210), Chapman and Foster (1969, pl. 1).

26-27. Little Eldorado Creek: Prindle and Katz (1909, p. 188, 190–191), Johnson (1910, p. 246), Byers (1957, p. 188, 210), Chapman and Foster (1969, pl. 1).

28. Last Chance Creek: Chapman and Foster (1969, pl. 1). Little Eldorado Creek: Prindle and Katz (1909, p. 188, 190–191), Johnson (1910, p. 246), Byers (1957, p. 188, 210), Chapman and Foster (1969, pl. 1). Louis Creek: Chapman and Foster (1969, pl. 1).

29-30. Chatanika River: Prindle and Katz (1909, p. 190–191), Chapman and Foster (1969, pl. 1).

31. Bedrock Creek: Joesting (1942, p. 32, 37), Byers (1957, p. 188, 210), Chapman and Foster (1969, pl. 1). Chatanika River: Prindle and Katz (1909, p. 190–191), Chapman and Foster (1969, pl. 1). Chatham Creek: Prindle (1910b, p. 226), Joesting (1942, p. 10–11, 32, 37), Joesting (1943, p. 9). Cleary Creek: Prindle (1906, p. 111, 119), Joesting (1942, p. 8, 10, 32, 37), Chapman and Foster (1969, pl. 1). Willow Creek: Chapman and Foster (1969, pl. 1). Wolf Creek: Prindle (1908a, p. 41–42).

32. Alder Creek: Chapin (1914a, p. 359), Smith (1942b, p. 39). Crane Gulch: Prindle and Katz (1913, p. 112–113). Deep Creek: Chapin (1914a, p. 359). Fairbanks Creek: Prindle and Katz (1913, p. 102), Joesting (1942, p. 11, 32, 37, 40), Byers (1957, p. 188, 210–211), Chapman and Foster (1969, pl. 1). Fish Creek: Chapman and Foster (1969, pl. 1). Walnut

Creek: Prindle and Katz (1913, p. 101), Ellsworth and Davenport (1913, p. 207–208).

33. Monte Cristo Creek (Pup): Eakin (1915b, p. 233).

34. Fish Creek: Prindle and Katz (1913, p. 102–103), Joesting (1942, p. 11, 32), Wedow, White, and others (1954, p. 1, 3).

35. Pearl Creek: Hill (1933, p. 71), Joesting (1943, p. 27), Smith (1942b, p. 39). Yellow Pup Creek: Chapman and Foster (1969, pl. 1).

36. Last Chance Creek: Eakin (1915b, p. 233).

37. Slippery Creek: Chapman and Foster (1969, pl. 1).

38. Fish Creek: Chapman and Foster (1969, pl. 1).

39. Unnamed creek: Chapman and Foster (1969, pl. 1).

40. Fish Creek: Chapman and Foster (1969, pl. 1).

41-42. Nugget Creek: Smith (1942b, p. 39), Chapman and Foster (1969, pl. 1).

43. Nugget Creek: Smith (1942b, p. 39), Chapman and Foster (1969, pl. 1). Smallwood Creek: Prindle (1908a, p. 46–47), Prindle and Katz (1913, p. 103).

44. Smallwood Creek: Prindle (1908a, p. 46–47), Prindle and Katz (1913, p. 103).

45. Kokomo Creek: Brooks (1923, p. 6, 29), Chapman and Foster (1969, pl. 1).

In general, the placers of the Fairbanks district were stream placers that had been buried by more recent alluvium and loess that choked creek valleys; a few deposits have been found on buried bedrock benches. Drift mining indicated that most of the rich placers were in the lowest parts of bedrock valleys that are much more symmetrical than those occupied by the present streams.

FORTYMILE DISTRICT

The Fortymile district (pl. 1, fig. 41) comprises the Alaska part of the Fortymile River drainage basin, the area drained by streams flowing southwest into the Tanana and Chisana Rivers and Scottie Creek from Tanacross to the Alaska-Yukon boundary, and the Alaskan parts of the basins of the western tributaries of the Sixtymile and Ladue Rivers.

The district consists of discontinuous groups of mountains that rise a few hundred feet above a fairly uniform plateau 3,000–5,000 feet above sea level. The upland is interrupted by the valleys of the Fortymile River and its forks and a few other streams.

Most of the district is underlain by Precambrian(?) and Paleozoic gneiss and schist, of both sedimentary and igneous origin, and minor crystalline limestone, and by Paleozoic, Mesozoic, and Tertiary felsic, mafic, and ultramafic igneous rocks (Foster, 1967, 1968, 1969a, b; Foster and Clark, 1970; Foster and Keith, 1968, 1969; Mertie, 1937b). Some of the felsic intrusive bodies, particularly those in the western and northern parts of the district, are batholiths of great areal extent. Tertiary detrital, volcanic, and volcaniclastic rocks occur locally.

Although the upland was not glaciated except for the summits of some of the highest mountains, loess, derived largely from proglacial flood plains, mantles much of the area. The district is in a zone of discontinuous permafrost. Much of the ground that was placer mined was permanently frozen.

Only a few lodes have been found in the Fortymile district (Berg and Cobb, 1967, p. 221–222, fig. 32); a little gold was mined from one near Chicken (Foster, 1969a, p. G28). The lodes consist of quartz veins in metamorphic rocks or, less commonly, contact-metamorphic deposits carrying gold, silver, lead, copper, zinc, antimony, and iron.

The Fortymile is one of the oldest placer districts in Alaska. Gold was discovered in the Yukon Territory near the mouth of the Fortymile River in the fall of 1886 and in Alaska on Franklin Creek (6, fig. 46) about a year later. In 1888, commercial gold placers were found at many other places, particularly in the valley of Walker Fork. By 1903, most of the productive deposits had been located. From the time of their discovery through 1961, placers in the Fortymile district were worked in every year, yielding a total of about 417,000 ounces of gold, 2 percent of the total placer-gold production of Alaska. Data on production in most years since 1961 are not available, even though a large dredge operated on Chicken Creek (10, fig. 46) through the end of the 1967 season. After this dredge shut down, only about half a dozen one-man operations were active.

The source of the gold in the placers was abundant small mineralized quartz veins in metamorphic rocks near contacts with felsic intrusive bodies. Most of the productive placers were near such contact zones, although many streams that cross contacts around plutons do not carry minable gold. Heavy minerals identified in concentrates include magnetite, ilmenite, hematite, barite, garnet, and pyrite and other sulfide minerals. Small amounts of scheelite were reported from Chicken Creek and its tributaries, Myers Fork (10, fig. 46) and Stonehouse Creek (9, fig. 46), the Mosquito (12, fig. 46) and Dennison (13, fig. 46) Forks of the Fortymile River, Wade Creek (16, fig. 46) and Fortyfive Pup (4, fig. 46). Cassiterite was found in samples from several of these streams. Cinnabar, for which no lode is apparent, is a constituent of concentrates from Dome (34, fig. 46), Franklin, Stonehouse, and Wade Creeks. A reconnaissance for radioactive deposits in 1949 (Wedow, White, and others, 1954) resulted in the discovery of monazite and allanite in material from Ben and Slate (33, fig. 41) and Ruby (34, fig. 41) Creeks. None of the heavy minerals except gold has been recovered commercially. Although much of the gold was fine, some was very coarse. A 25¼-ounce nugget was found on Wade Creek, and nuggets weighing several ounces were not uncommon.

Both streams and bench placers have been mined in the Fortymile district. The most productive benches were those along upper Chicken Creek and its tributaries and at the head of and along the west side of neighboring Lost Chicken Creek (11, fig. 46). Mining, particularly in the early days, was by drifting in frozen parts of the bench gravels, but hydraulic and nonfloat methods also were used. An auriferous bench deposit that extends for several miles along the north wall of the valley of Dome Creek (34, fig. 46) was mined by hydraulic methods in several places. Little Miller Creek (34, fig. 46), a small tributary of Dome Creek, cut through the bench and reconcentrated gold from the gravels, forming a very rich stream deposit that was mined out in the

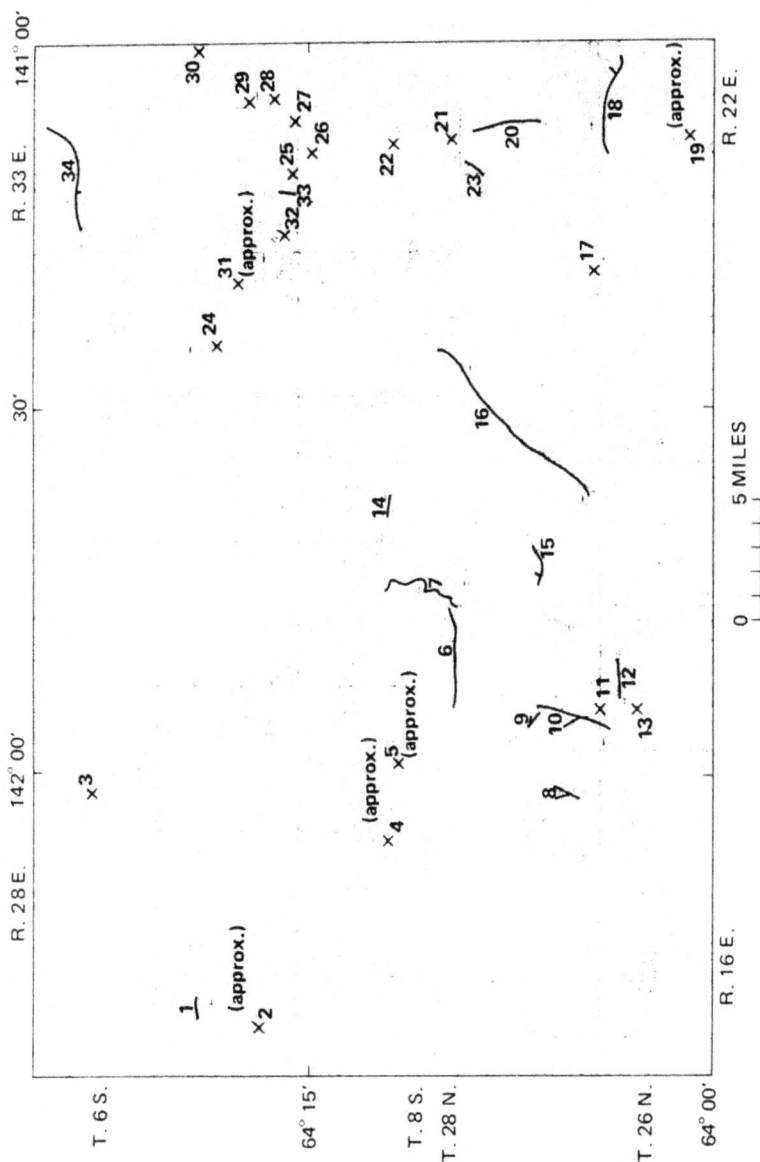

FIGURE 46.—Placer deposits in the Fortymile River area.

FIGURE 46.—Continued.

1. Hutchinson Creek: Prindle (1913b, p. 80). Montana Creek: Prindle (1905, p. 52), Prindle (1913b, p. 80).
2. Confederate Creek: Prindle (1905, p. 52).
3. Fortymile River, North Fork: Prindle (1905, p. 51–52), Mertie (1937b, p. 245).
4. Fortyfive Pup: Mertie (1938a, p. 185), Joesting (1943, p. 19–20, 28).
5. Buckskin Creek: Mertie (1938a, p. 185).
6. Franklin Creek: Prindle (1908b, p. 194), Mertie (1938a, 181–193).
7. Fortymile River, South Fork: Mertie (1938a, p. 185–186).
8. Ingle Creek: Mertie (1938a, p. 171, 173, 179). Lilliwig Creek: Mertie (1938a, p. 180).
9. Irene Gulch: Prindle (1909, p. 39–40). Stonehouse Creek: Mertie (1938a, p. 177), Joesting (1943, p. 20).
10. Chicken Creek: Mertie (1938a, p. 170–175), Joesting (1943, p. 20), Foster and Keith (1969, p. 17–18). Myers Fork: Mertie (1938a, p. 175–177), Joesting (1943, p. 20).
11. Lost Chicken Creek: Mertie (1938a, p. 177–179).
12. Fortymile River, Mosquito Fork: Mertie (1938a, p. 180–181), Wedow, White, and others (1954, p. 10–12).
13. Fortymile River, Dennison Fork: Porter (1912, p. 214), White, Nelson, and Matzko (1963, p. 81).
14. Uhler Creek: Foster (1969a, fig. 13).
15. Napoleon Creek: Mertie (1938a, p. 183–185), Foster 1969a, p. G28).

16. Gilliland Creek: Mertie (1938a, p. 168–169). Wade Creek: Martin (1919, p. 21), Mertie (1938a, p. 163–170), Foster and Keith (1969, p. 23).
17. Walker Fork: Prindle (1909, p. 36), Mertie (1938a, p. 160).
18. Davis Creek: Mertie (1938a, p. 157–159). Poker Creek: Mertie (1938a, p. 157–160). Walker Fork: Mertie (1938a, p. 157–163).
19. Cherry Creek: Prindle (1905, p. 43–44).
20. Camp Creek: Brooks (1916a, p. 62). Canyon Creek: Mertie (1938a, p. 187), Smith (1941b, p. 48). Woods Creek: Prindle (1908b, p. 197).
21-22. Canyon Creek: Ellsworth and Parker (1911, p. 169).
23. Squaw Gulch: Prindle (1909, p. 41–42).
24. Fortymile River: Prindle (1909, p. 43), Mertie (1938a, p. 186–188).
25-26. Fortymile River: Mertie (1938a, p. 186–188).
27. Fortymile River: Prindle (1909, p. 43), Mertie (1938a, p. 186–188).
28-29. Fortymile River: Mertie (1938a, p. 186–188).
30. Fortymile River: Mertie (1938a, p. 185).
31. Flat Creek: Brooks (1916a, p. 62).
32. Fortymile River: Ellsworth and Parker (1911, p. 170), Mertie (1938a, p. 186–188). Twin Creek: Prindle (1908b, p. 197).
33. Nugget Gulch: Prindle (1905, p. 52).
34. Dome Creek: Mertie (1938a, p. 188–190). Little Miller Creek: Mertie (1938a, p 188–189).

middle 1890's. Rich pay streaks were found in a bench on Napoleon Creek (15, fig. 46).

Although practically every kind of placer-mining method has been used in the Fortymile district, most of the production was from dredges. The first dredges were installed on Walker Fork near the mouth of Twelvemile Creek (17, fig. 46) and between the mouths of Davis and Poker Creeks (18, fig. 46) in 1907. One of these was later moved to the South Fork of the Fortymile near the mouth of Uhler Creek (7, fig. 46). In 1934, a dredge was built on the upper part of Walker Fork near Boundary (18, fig. 46) and was operated for several years before being abandoned near the mouth of Cherry Creek. In 1936, a dredge, fitted out with the machinery from the one that had operated near Franklin for many years, began digging on lower Wade Creek (16, fig. 46). It now lies abandoned about half a mile below the mouth of Ophelia Creek. Dredges were operated successfully in other parts of the district, including Canyon Creek (20, fig. 46), the Fortymile River near the international boundary (30, fig. 46), Poker Creek (18, fig. 46), the Mosquito Fork of the Fortymile at Atwater Bar (12, fig. 46), and, most recently, Chicken Creek. Dredging was unsuccessful at the Kink, an artificially cut off meander of the North Fork of the Fortymile River (3, fig. 46), although gold is present.

Mining by methods other than dredging has been carried on at Ben Creek (33, fig. 41), Gold Run (35, fig. 41), at most other localities shown in figure 46, and at many other places that are not well enough described to locate accurately. Of particular interest, though their total production probably was not statistically very important, are many bars of the Fortymile River such as Claghorn (29, fig. 46) and Bonanza (25, fig. 46) Bars, where gravel is either very thin or missing entirely. Cleavage in the bedrock is nearly vertical and acts as natural riffles to concentrate gold which can then be recovered with simple equipment. These placers are self renewing, as each flood deposits new material from upstream, allowing them to be mined many times.

GOODPASTER DISTRICT

The Goodpaster district (pl. 1, fig. 41) is the area drained by northern tributaries of the Tanana River between Big Delta and Tanacross. In physiography and geology it is similar to the adjoining Fortymile district.

The only lode production from the Goodpaster district consisted of a small quantity of gold ore mined from free gold- and sulfide-bearing quartz veins on two claims near Tibbs Creek (37, fig. 41). Antimony and molybdenum minerals have been found there and at one other place in the district (Berg and Cobb, 1967, p. 222–223, fig. 32).

Prospecting before World War I in the Goodpaster and neighboring valleys resulted in reports of gold on several streams, but descriptions were so vague that the sites of the purported discoveries cannot be determined. Gold on Kenyon Creek, said to be in the Healy River area, was reported to be in very deep gravels resting on schist (Brooks, 1916a, p. 50–51). A little gold, including a nugget worth $9, was recovered from Michigan Creek (38, fig. 41) in 1916, but no mining has been reported since then. Some placer-gold prospecting and mining has been reported on Tibbs Creek and its tributaries (R. M. Chapman, written commun., 1953). Placer gold production in the Goodpaster district was so small that it does not appear in tables of production data; it probably was not more than a few hundred ounces.

HOT SPRINGS DISTRICT

The Hot Springs district (pl. 1, fig. 47) is the area drained by southern tributaries of the Yukon River between Kallands and (but not including) Fish Creek. Tributaries of the Tanana River as far upstream as Dugan Creek are in the district.

The part of the district north of the Yukon River consists of several even-topped ridges 1,600–2,500 feet in elevation and a few more rugged isolated mountains with summits below 5,000 feet. Wide valleys merge with a lowland, much of which is less than 500 feet above sea level, along the Tanana River. The southern part of the district is made up of groups of hills 2,000–3,200 feet in elevation that slope northward and northeastward to the lowland and are separated by broad valleys.

North of the Tanana River, Paleozoic chert, limestone, and clastic rocks, many somewhat sheared and metamorphosed, are the oldest units. They are overlain by Cretaceous argillaceous rocks, graywacke, and sandstone that in much of the area were metamorphosed to slate, phyllite, and quartzite. The youngest consolidated bedded rocks are Tertiary continental conglomeratic beds in a small area near the Yukon River about 20 miles upstream from Tanana. Bodies of mafic rocks, now largely serpentinized, intruded the bedded rocks near Tofty. Felsic plutons of probable Tertiary age that range in composition from monzonite to biotite granite make up several of the mountains and high parts of ridges (Eakin, 1913a; Mertie, 1937b). South of the Tanana and Yukon Rivers, clastic, carbonate, and volcanic rocks, some of Paleozoic and some of Mesozoic and probable Tertiary age, were

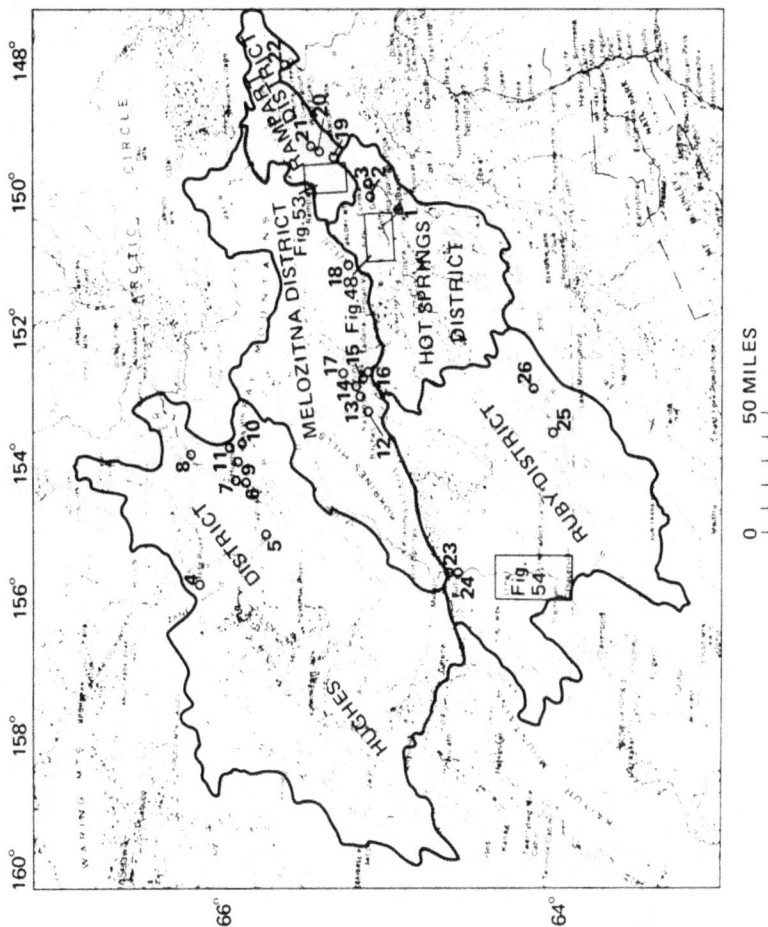

FIGURE 47.—Placer deposits in the Hot Springs, Hughes, Melozitna, Rampart, and Ruby districts.

FIGURE 47.—Continued.

Hot Springs district

1. Karshner Creek: Waters (1934, p. 242).
2. Chicago Creek: Hess (1908, p. 92). Glen Creek (Gulch): Mertie (1934, p. 203). Wayland (1961, p. 396). Gold Run: Mertie (1934, p. 200-201). Omega Creek: Mertie (1934, p. 203-204). Waters (1934, p. 238). Rhode Island Creek: Mertie (1934, p. 199-201). Seattle Creek: Mertie (1934, p. 200). Shirley Bar (Bench): Mertie (1934, p. 193, 201-202). Waters (1934, p. 237-238). Thanksgiving Creek: Hess (1908, p. 82-83, 92-93, 97-98).
3. Eureka Creek: Mertie (1934, p. 166, 192-195). Wayland (1961, p. 396). McCaskey Bar: Mertie (1934, p. 193, 198-199). Waters (1934, p. 237). Boothby Creek, Doric Creek, Jordan Bar, Last Bench, Pioneer Creek, Seattle Jr. Creek, Skookum Seattle Bar, What Cheer Bar: Mertie (1934, p. 195-198, 201), Waters (1934, p. 236-237).

Hughes district

4. Bear Creek (Hogatza, Hog River): Miller and Ferrians (1968, p. 9), unpub. data.
5. Batza Slough: Miller and Ferrians (1968, p. 6).
6. Florence Bar: Smith (1913, p. 142).

Hughes district—Continued

7. Hughes Bar: Smith (1918, p. 142), Reed (1938, p. 163).
8. Red Mountain: Eakin (1914c, p. 388), Reed (1938, p. 163).
9. Pocahontas Creek: Eakin (1916, p. 83).
10. Utopia Creek: Smith (1942b, p. 47). Miller and Ferrians (1968, p. 3, 5).
11. Black Creek: Eakin, (1916, p. 83-84). Felix Fork (Creek): Smith (1913, p. 143), Martin (1919, p. 39). Indian River (Creek): Eakin (1916, p. 82-84). U.S. Geological Survey (1965, p. A102). Patton and Miller (1966). Snyder Creek: Smith (1913, p. 143).

Meloxitna district

12. Mason Creek: Chapman, Coats, and Payne (1963, p. 33-35).
13. Illinois Creek: Eakin (1913a, p. 34), Chapman, Coats, and Payne (1963, p. 16).
14. Golden Creek: Chapman, Coats, and Payne (1963, p. 16).
15. American Gulch: Eakin (1912, p. 282).
16. Grant Creek: Smith (1932, p. 68), Chapman, Coats, and Payne (1963, p. 16). Grant (Lynx) Creek: Eakin (1912, p. 281-282).
17. Tozimoran Creek: Chapman, Coats, and Payne (1963, p. 14-26).

Meloxitna district—Continued

18. Bonanza Creek: Chapman, Coats, and Payne (1963, p. 5, 9). Homestake Creek: Brooks (1911, p. 184). Morelock Creek: Chapman, Coats, and Payne (1963, p. 5-18).

Rampart district

19. Quail Creek: Mertie (1934, p. 189-191), Waters (1934, p. 235). Troublesome Creek: Prindle and Katz (1913, p. 146).
20. Gunnison Creek: Mertie (1934, p. 192), Waters (1934, p. 235-236).
21. Troublesome Creek: Mertie (1934, p. 192), Waters (1934, p. 236).
22. Willow Creek: Smith (1926, p. 14).

Ruby district

23. Ruby Creek: Mertie and Harrington (1916, p. 260), Mertie (1936, p. 144).
24. Big Creek: Mertie (1936, p. 144-145), Chapman, Coats, and Payne (1963, p. 37, 40, 44-46). Cox Pup (Gulch): Chapman, Coats, and Payne (1963, p. 40, 44-46). Glacier Creek: White and Stevens (1963, p. 8), Chapman, Coats, and Payne (1963, p. 37, 46-47).
25. American Creek: Eakin (1918, p. 51).
26. Baker Creek: Eakin (1918, p. 51).

intruded by at least two granitic plutons (Eakin, 1918). Practically all major lowland areas in the district are covered by Quaternary glaciofluvial material derived from the Alaska Range to the south, alluvium, and loess. Much of this material is perennially frozen and has a high content of organic material. No part of the Hot Springs district has been glaciated.

None of the few lode deposits in the district has been productive. A shear zone on the summit of Manley Hot Springs Dome contains argentiferous galena and minor amounts of gold and various sulfide minerals and their alteration products (Berg and Cobb, 1967, p. 223, fig. 34). Chromite grains are disseminated in a serpentinized mafic rock on Serpentine Ridge north of the head of Woodchopper Creek (R. M. Moxham, written commun., Nov. 4, 1949).

The original discovery of gold in the Hot Springs district was on Eureka Creek (3, fig. 47) in 1898 by a group of men from New England nicknamed the "Boston Boys." The next spring, a stampede from Rampart began, resulting in the establishment of the Eureka area as a recognized mining community. Another center of mining in the district developed at Tofty, where gold was found on Tofty Gulch (9, fig. 48) in the winter of 1906-7. Through 1960, about 450,000 fine ounces of gold, or about 2.2 percent of the State total, was taken from placers in the district. In 1961, the last year for which data are available, five mines are reported to have produced 1,062 ounces of gold and a little alloyed silver. Production since then has probably been on about the same scale. In 1968, seven nonfloat operations, the largest of which employed five men, were reported. Several hundred tons of cassiterite concentrates have been produced as a byproduct of gold mining.

Definite sources of the valuable heavy minerals in the placer deposits of the district have not been found, although much effort and capital have been expended searching for them. The thick frozen silt and moss cover makes prospecting difficult; data on bedrock in much of the district can come only from drill holes, prospect shafts, or the study of ephemeral exposures in placer mines. Wayland (1961, p. 398) estimated that near Tofty "probably less than 1 in 300 square feet of bedrock of this 2-mile-wide tin belt had been seen or tested by 1941." An even smaller proportion would probably apply for the area near Eureka.

The present drainage pattern in the Eureka and Tofty areas reflects, but does not follow, an older pattern that was part of a mature land surface with a thick residual cover and many bedrock terraces on which gravel deposits rested. The old erosion cycle was interrupted, probably because aggradation in the Tanana

Valley raised the local base level, and the valleys were filled with alluvium and loess. The placers that have been mined are nearly all buried bench gravels (locally called bars) on old terraces, or buried stream deposits that were enriched where small tributaries cut through older bench gravels. In the Eureka area such terrace deposits extend parallel to upper Eureka and Pioneer (3, fig. 47) Creeks and from Thanksgiving Creek (2, fig. 47) to and beyond Rhode Island Creek (2, fig. 47) nearly to Eureka. Another, McCaskey Bar (3, fig. 47), is preserved on a spur between Pioneer and Kentucky Creeks. Deposits in the Tofty area, commonly called the Tofty tin belt because of the large amounts of cassiterite in the placers, are similar to those near Eureka, with bench gravels extending in a zone up to 2 miles wide from Woodchopper Creek (5, fig. 48) to Killarney Creek (15, fig. 48). A smaller bench follows Quartz Creek (17, fig. 48) but is lower in gold content and was not mined extensively. American Creek (1, fig. 48) and its tributary New York Gulch (2, fig. 48) have not been studied in detail, but the deposits probably are similar to those near Eureka and Tofty. The shape of the area mined on American Creek and its orientation suggest that much of it may be a stream placer, possibly derived only in part from an older bench deposit. On Boulder Creek (3, fig. 48) most of the mining was on a bench about 1,000 feet north of the present stream.

The Tofty tin belt presents interesting problems, as no bedrock occurrence of either cassiterite or gold has been found anywhere in the area. Cassiterite is always associated with brown tourmaline and quartz in the stream-gravel concentrates, in many instances in the same rock fragments. In one piece of subangular phyllite from Miller Gulch (7, fig. 48), for example, fine brown tourmaline replaced all formerly micaceous layers, and dense tourmaline filled crosscutting fractures. Yellow fluorite also occurs in some pebbles that are composed mainly of cassiterite. Granite magma is almost universally accepted as the ultimate source of practically all cassiterite, although the mineral itself may be a constituent, usually accompanied by tourmaline and fluorite, of veins relatively far from the parent body. The only body of granite exposed near the Tofty tin belt is at Manley Hot Springs Dome. Between this body and the tin belt, however, no sign of a typical tin-district mineral assemblage has been found in either bedrock or stream sediments, which strongly suggests that the cassiterite could not have come from the granite at Manley Hot Springs Dome. Mertie (1934, p. 207–208), Waters (1934, p. 245–246), and Wayland (1961, p. 398–403) gave data supporting a

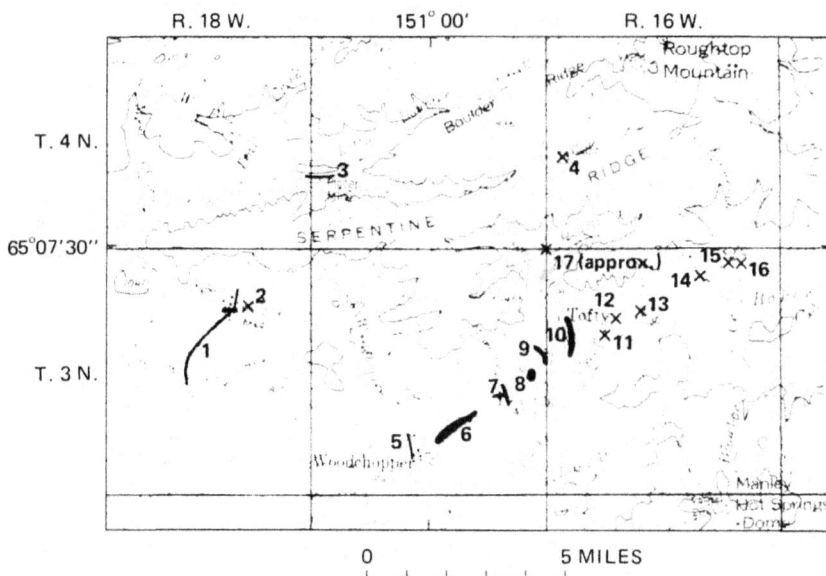

FIGURE 48.—Placer deposits in the Tofty area.

1. American Creek: Mertie (1934, p. 213–214), Waters (1934, p. 241).
2. New York Gulch: Mertie (1934, p. 214), Waters (1934, p 241).
3–4. Boulder Creek: Mertie (1934, p. 214–215), Waters (1934, p. 241), Moxham (1954, p. 6).
5. Patterson Creek: Thomas (1957, p. 7). Woodchopper Creek: Thomas (1957, p. 7), Wayland (1961, p. 372, 388–389, 392–394, 399, 402–403, 408).
6. Deep Creek, Hokeley Gulch, Innesvale Gulch: Thomas (1957, p. 7, 45), Wayland (1961, p. 367, 374, 385–392, 397, 408, 410). Patterson Creek: Thomas (1957, p. 7).
7. Miller Gulch: Moxham (1954, p. 5–6), Thomas (1957, p. 6–8), Wayland (1961, p. 370, 373–374, 384–385, 408–410). Patterson Creek: Thomas (1957, p. 7.)
8. Idaho Gulch: Moxham (1954, p. 5), Thomas (1957, p. 6–8), Wayland (1961, p. 372, 379, 382–384, 397–398). Patterson Creek: Thomas (1957, p. 7).
9. Patterson Creek: Thomas (1957, p. 7). Tofty Gulch: Moxham (1954, p. 5–6),

Thomas (1957, p. 7–8), Wayland (1961, p. 374, 379–382, 403, 405–411).
10. Patterson Creek: Thomas (1957, p. 7). Sullivan Creek: Wayland (1961, p. 367, 372, 374, 377–379, 396, 409–411).
11. Harter Gulch: Thomas (1957, p. 48), Wayland (1961, p. 376). Patterson Creek: Thomas (1957, p. 7).
12. Dalton Gulch: Thomas (1957, p. 7–8, 48), Wayland (1961, p. 376). Patterson Creek: Thomas (1957, p. 7).
13. Cache Creek: Mertie (1934, p. 212), Waters (1934, p. 240), Wayland (1961, p. 365–375, 397). Ferguson Draw: Wayland (1961, p. 374–375). Patterson Creek: Thomas (1957, p. 71).
14. Gold Basin Creek: Wayland (1961, p. 394–395, 399).
15. Killarney Creek: Wayland (1961, p. 372, 394–399).
16. Cooney Creek: Smith (1939b, p. 50), Wayland (1961, p. 294).
17. Quartz Creek: Mertie (1934, p. 212–213), Waters (1934, p. 240), Wayland (1961, p. 396). Sullivan Creek: Wayland (1961, p. 367, 372, 374, 377–379, 396, 403–411).

hypothesis of origin involving deposition of cassiterite in veins along a shear zone in a body of rock spatially above the present tin belt, later erosion by small streams of low gradient, and the concentration of cassiterite and other heavy minerals as a residual deposit that was later buried. Wayland pointed out that the cassiterite and gold need not have come originally from the same source.

Concentrates from placers in the Hot Springs district have been extensively studied in investigations mainly designed to determine the source of the cassiterite in the Tofty tin belt. As a result, more minerals have been identified than from most mining districts in Alaska. Outside of the tin belt, a little cassiterite has been found near Eureka, where it may have been introduced by pranksters, and in concentrate from a bench on Quartz Creek about 2 miles from the tin belt. Chromite, picotite (chrome spinel), ilmenite, and magnetite have been derived from serpentinized ultramafic bodies found in several deposits; more than 10 percent of the concentrates at Deep Creek (6, fig. 48) downstream from Hokeley Gulch are chromite. Sulfide minerals are less common than in most Alaskan placer districts, but traces of cinnabar have been reported from a few places near Eureka. Rare-earth minerals, including monazite, aeschynite, and xenotime are fairly widespread but not concentrated into minable deposits. Scheelite was identified in concentrates from Omega (2, fig. 47) and Pioneer Creeks and native copper in material from Sullivan Creek (10, fig. 48).

Mining in the Hot Springs district has always been hampered by a shortage of water and by low gradients that make it difficult to achieve a good head on what water is available. Tailings disposal is also difficult, and it has been necessary to use either excessively long strings of sluice boxes set at inefficiently low grades or very long and deep bedrock drains. As a result, much of the mining was by drifting on bedrock from shafts in frozen muck and gravel. Some of these operations were extensive and relatively efficient. Most of the large plants used steam scrapers, bulldozers, and other mechanical means for getting gravel to sluice boxes. A dredge that operated for several years on American Creek was the only one in the district. All open-cut and dredge operations required thawing and stripping of barren frozen ground in advance of actual mining. Most stripping was by ground sluicing off increments that thawed naturally, a slow process at best; the technologically more advanced methods developed in the Fairbanks and Nome districts were not used here. The top few feet of weathered bedrock generally had to be mined with the overlying gravel to insure good recovery, as much fine gold had worked down into cracks. Miners usually took up bedrock until they found relatively fresh pyrite for experience had shown that further excavation would be unprofitable.

HUGHES DISTRICT

The Hughes district (pl. 1, fig. 47) includes the area drained by the Koyukuk River and its tributaries below the Kanuti River and by northern tributaries of the Yukon River between Koyukuk and Melozi.

The central part of the district is a swampy, lake-studded lowland generally less than 200 feet above sea level through which the Koyukuk River and its tributaries meander. Low mountains and uplands that fringe the lowlands are commonly 2,000–3,000 feet in elevation, although Indian Mountain, east of Hughes, is more than 4,200 feet high. Several broad passes and ill-defined drainage divides are below 1,000 feet.

The mountains are underlain mainly by Jurassic and Cretaceous marine and nonmarine clastic and volcanic rocks that were intruded by large granitic and monzonitic plutons and small bodies of latitic hypabyssal rock of Late Cretaceous age (Cass, 1959c–1959e; Patton, 1966; Patton and Miller, 1966; Patton and others, 1968). Lowlands near the major streams are underlain by thick alluvial and glaciofluvial deposits mantled by eolian deposits that include large areas of active sand dunes. Pleistocene ice from the Brooks Range covered the northern part of the district and parts of the Zane Hills, Purcell Mountains, and Indian Mountains, but most of the area was not glaciated and there are no modern glaciers in it. Except for a permafrost-free area in the Zane Hills west of Hogatza (Hog River), the district is underlain by continuous permafrost.

The only lode deposits known in the district are near Indian Mountain, where small amounts of zinc and copper sulfides occur near two granitic stocks (Miller and Ferrians, 1968, p. 5–6). Joesting (1942, p. 29) reported that molybdenite had been found in bedrock exposed during placer mining on Indian River (11, fig. 47). Near the head of Utopia Creek (10, fig. 47), massive barite boulders, for which a bedrock source was not found, contain tetrahedrite, galena, and sphalerite. At Batza Slough (5, fig. 47), an angular float block of altered silicified rock was found to contain disseminated cerrusite, galena, and malachite and, as determined by analysis, 3 ounces of silver per ton. Stream-sediment samples collected in the southern Zane Hills during a geochemical reconnaissance (Miller and Ferrians, 1968) contained anomalous concentrations of several metals, including uranium and thorium.

Placer gold was discovered late in the 19th century on bars of the Koyukuk River, of which Florence and Hughes Bars (6, 7, fig. 47) were the most noteworthy. Most later mining was at Bear

Creek (4, fig. 47) and on streams that drain Indian Mountain east of Hughes. No accurate estimate of the amount of gold produced from the district can be made, as in most years production was incorporated with that from other districts. The total production definitely allocatable to the Hughes district through 1960 is about 8,500 fine ounces, a figure obviously much too small in view of the extensive mining on Bear and Utopia Creeks and the Indian River. Except for the bars of the Koyukuk River, all the gold came from streams that drain contact zones around granitic plutons near Indian Mountain and in the southern Zane Hills. Utopia Creek and the upper Indian River and its headwater tributaries were the most important sites of mining until a large dredge was installed on Bear Creek in 1957. This dredge was one of the few in Alaska that was still operating in 1969, and while no production data have been made public, it undoubtedly has accounted for a significant part of the recent gold production of the State. It is known, on the basis of fragmentary unpublished records, that cassiterite and metals of the platinum group were identified in samples collected from Bear Creek in the early 1920's. Deposits on Utopia Creek and the Indian River were first worked by hand methods, but most production was probably from larger scale nonfloat operations that ceased on Utopia Creek in 1952 and on the Indian River in 1961. Other than the dredge on Bear Creek, no mining activity in the Hughes district was reported in 1968.

IDITAROD DISTRICT

The Iditarod district (pl. 1, fig. 37) comprises the area drained by eastern tributaries of the Yukon River between and including Paimut and Shageluk Sloughs, by southern tributaries of the Innoko River between Holikachuk and Dishkakat, and by tributaries of the Innoko below Holikachuk.

The southeastern part of the area consists of rounded northeast-trending ridges, generally less than 1,500 feet in altitude, that are surmounted by a few isolated rugged mountains, most of which are on the Yukon-Kuskokwim divide and all of which are lower than 3,000 feet. The rest of the district is low (at most only a few hundred feet above sea level), lake speckled, swampy, and crossed by widely meandering tributaries and sloughs of the Yukon River.

The upland part of the Iditarod district is underlain mainly by Paleozoic and Mesozoic metamorphic, clastic, and volcanic rocks. Tertiary mafic and felsic rocks form sills and other small intrusive bodies and, near Flat, small Tertiary monzonite stocks, which are possibly apophyses of a larger incompletely unroofed pluton, metamorphosed the surrounding Cretaceous clastic rocks (Cady and others, 1955; Eakin, 1913b, 1914a; Hoare and Coonrad, 1959b; Mertie and Harrington, 1924). Much of the highland area is mantled by a blanket of weathered residual material, colluvium, and silt. The valleys and lowlands near the Yukon and Innoko Rivers are covered by thick masses of alluvium. As the mountains are not high enough for Pleistocene cirques to have formed and ice from glaciers in adjoining areas did not invade the Iditarod district, there are no glacial deposits. Most of the district is in a zone of moderately thick to thin permafrost, though the lower Iditarod and Innoko Valleys are underlain by isolated masses of perennially frozen ground rather than continuous permafrost.

Lodes in the Iditarod district contain cinnabar, stibnite, gold, silver, and lead and zinc minerals (Berg and Cobb, 1967, p. 226–227, fig. 29; Maloney, 1962). The DeCourcy Mountain mine (near locality 2, fig. 37), the source of more than 1,200 flasks of mercury, is the major Alaskan mercury mine outside the Kuskokwim River region. In the Flat area (fig. 49), many quartz and calcite veins carrying stibnite, cinnabar, other sulfide minerals, and gold occur in two monzonite stocks and in contact-metamorphic aureoles in the surrounding Cretaceous bedded rocks. In the 1920's and 1930's, gold, silver, lead, and zinc were mined from a lode on the Golden Horn property a short distance south of Discovery (fig. 49) (Maloney, 1962, p. 4).

Gold was discovered on Otter Creek (1, fig. 49) on Christmas Day 1908 by prospectors from the Ophir area of the adjoining Innoko district, but little work was done until 1910, when more than 2,000 people and much mining machinery arrived and several towns sprang up in the area. Many of the early stampeders soon left, but enough stayed and brought in more equipment, including the first dredge in the district (installed on Flat Creek (1, fig. 49) in 1912), to recover more than 130,000 fine ounces of gold in 1912, the most productive single year for the district. Placer mining has been reported for every year since 1912. Total production through 1966 was 1,329,404 fine ounces of gold, more than 6 percent of Alaska's placer-gold production (Kimball, 1969b, p. 3). In 1968, a small dredge and three hydraulic or nonfloat operations were active, but the amount of gold recovered is not known. Almost all the mining in the district was within an area of about 60 square miles, and all was within 9 miles of Flat. The only other place where gold may have been mined commercially is Little Creek (2, fig. 37), where gold and cinnabar were found. The amount of gold recovered there is not known, but some of

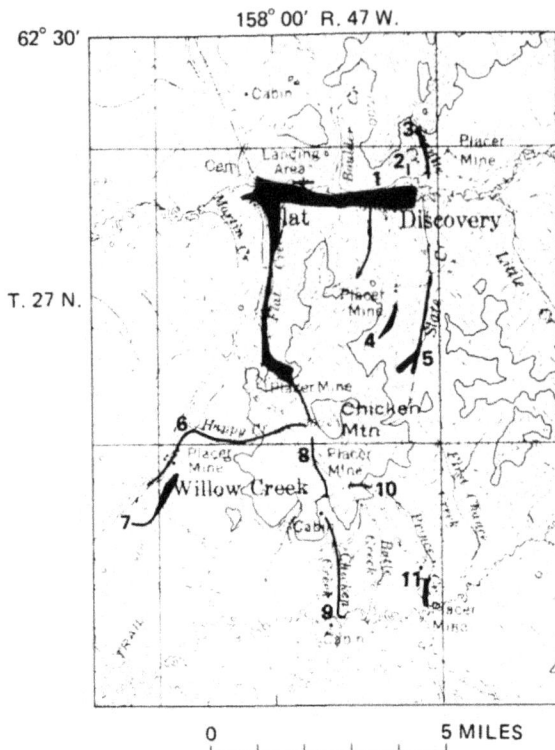

FIGURE 49.—Placer deposits in the Flat area.

1. Black Creek (Gulch): Maddren (1915, p. 286), Joesting (1942, p. 35). Flat Creek: Mertie and Harrington (1916, p. 255-258), Smith (1917, p. 152), Mertie (1936, p. 201-208, 243), Maloney (1962, p. 17). Glen Gulch (Creek): Mertie (1936, p. 218-219), Maloney (1962, p. 8-9, 17-20), Otter Creek: Mertie (1936, p. 216-220), Maloney (1962, p. 8, 17, 21).

2. Malamute Pup (Creek, Gulch): Maloney (1962, p. 17, 21).

3. Granite Creek: Mertie (1936, p. 220-221), this rept.

4-5. Slate Creek: Mertie (1936, p. 222-223), Maloney (1962, p. 17).

6. Happy Creek: Mertie (1936, p. 211-213), White and Killeen (1953, p. 9), Maloney (1962, p. 8, 17-18, 21-22). Willow Creek: Mertie (1936, p. 208-211).

7. Gold Creek (Run): Mertie (1936, p. 209). Willow Creek: Mertie (1936, p. 208-211).

8-9. Chicken Creek: Mertie (1936, p. 213-215), Maloney (1962, p. 16-17, 21).

10-11. Prince Creek: Mertie (1936, p. 216), Maloney (1962, p. 8, 18, 22).

the deposits were reported to contain gold worth a few cents to a dollar per square foot of bedrock cleaned.

The sources of placer gold in the Flat area are veins in the monzonite stocks and surrounding contact aureoles on Chicken Mountain and near Discovery. Weathered deposits in the area common to the heads of Flat (1, fig. 49), Happy (6, fig. 49), and Chicken (9, fig. 49) Creeks were rich enough to mine as residual placers. The residual placers grade downhill into stream placers, most of which are in deposits of the modern creeks, although a few are on terraces that merge downstream into the present streambeds.

Although many placer-mining methods have been employed in the Flat area, most production has been from operations that used dredges or large-scale nonfloat plants. Where additional water was needed, it was usually brought from farther upstream on the creek being mined rather than from great distances as was common in the Seward Peninsula region. The most productive creeks were Otter and Flat Creeks, both of which were dredged through most of their minable lengths even though some stretches had been worked earlier by other methods.

Many heavy minerals other than gold have been identified in placer concentrates from the Flat area. Cinnabar occurs in almost every stream; some may have been saved and retorted to produce quicksilver for local use. Chromite, probably derived from small ultramafic dikes, some of which have been found in placer pits, was reported from Granite Creek (3, fig. 49) and Malamute Pup (2, fig. 49) north of Otter Creek, from Otter Creek, and from Happy and Chicken Creeks. Scheelite, identified in samples from all of these streams except Granite Creek, probably came from veins in or around the monzonite stocks. Other heavy minerals from creeks in the area include magnetite, cassiterite, zircon, stibnite, arsenopyrite, galena, monazite, and allanite.

INNOKO DISTRICT

The Innoko district (pl. 1, fig. 37) is the area drained by the Innoko River and its tributaries above Dishkakat, excluding the Poorman area (which is in the Ruby district).

Most of the district is a rolling upland, generally less than 2,000 feet in elevation, above which rise several isolated rugged masses with summits 3,500 to more than 4,200 feet above sea level. Many of the streams flow northeast or southwest and the drainage pattern shows evidence of a complex history. The major valleys are swampy and are dotted with small lakes of probable thermokarst origin.

The upland part of the district is underlain mainly by Cretaceous clastic rocks, some of which now exhibit slaty cleavage. Older rocks include Paleozoic and Mesozoic metasedimentary and metavolcanic rocks and chert. Cretaceous and Tertiary igneous rocks include basaltic lava flows and tuffs, dikes and other small intrusive bodies ranging in composition from rhyolite to pyroxenite, and monzonitic plutons that in places metamorphosed the inclosing rocks to hornfels (Eakin, 1914a; Harrington, 1919a;

Maddren, 1910; Mertie and Harrington, 1916, 1924; Mertie, 1936).
Surficial deposits include silt, largely wind deposited, and colluvium and residual weathered material that hide bedrock in much
of the upland area. The higher isolated mountain groups supported Pleistocene cirque glaciers, but most of the district was
ice free. Permafrost underlies most of the area but is discontinuous beneath parts of the Innoko and Dishna Valleys and some
of the higher hills and mountains.

Lodes in the Innoko district contain stibnite, cinnabar, and gold
(Berg and Cobb, 1967, p. 227–228, fig. 39). The only production
was a little gold from a quartz vein next to a felsic dike about 10
miles southeast of Ophir. Similar lodes are probably hidden beneath the widespread unconsolidated deposits.

Prospectors passed through parts of what is now the Innoko
district before 1900, but gold was not discovered until 1906, when
colors were found on bars of the Innoko River. Later in the same
year, rich placers were located on Ganes Creek (11, fig. 50) about
10 miles above its mouth. Early in 1907, a stampede brought more
than a thousand men to the Ophir area and most of the creeks
were staked, although little gold was discovered until after most
of the stampeders had left. The area near Tolstoi (3–5, fig. 37)
was the scene of prospecting at this time, but valuable deposits
were not found until the winter of 1915–16, when production from
Boob Creek (3, fig. 37) began. Meanwhile, in 1912, the placers of
the Cripple Creek area (6–7, fig. 37) had been discovered. Development was slow and it was many years before substantial production was achieved. Through 1961 the Innoko district was the
source of somewhat more than 540,000 fine ounces of placer gold,
about 2.7 percent of the total for the State. Mining was reported
in every year through 1967, but no production figures are available for years after 1961. Seven placer mines, employing a total
of 18 men, were operating in 1967, but no activity was reported
in 1968. An unknown, but certainly small, amount of platinum
was recovered as a byproduct of gold mining on Boob Creek.

Most of the placer mining in the district was near the original
discovery on Ganes Creek. The productive streams in this area
(fig. 50) drain hills 1,700–2,000 feet in elevation between the
upper Innoko River and Beaver Creek. The highland source area
is underlain largely by Cretaceous slate. Many small dacite and
andesite dikes, some mineralized with pyrite, closely follow the
cleavage of the slate. Maddren (1910, p. 64) considered quartz
veins and these dikes to be the probable source of the placer gold.
Mertie (1934, p. 174–175), however, demurred, as vein quartz is
not a common constituent of the stream gravels. He stated: "It

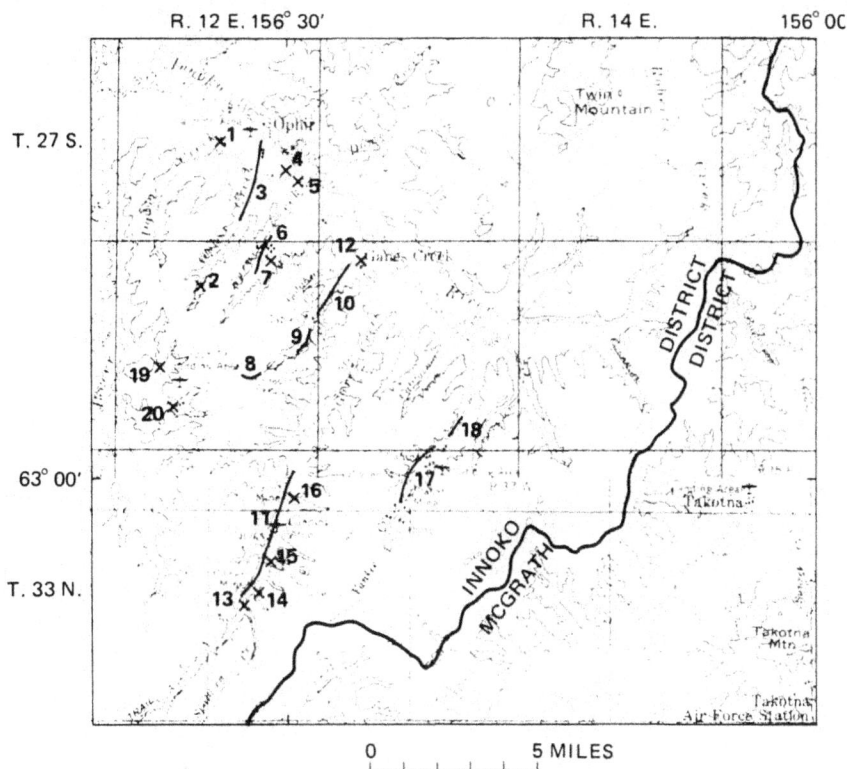

FIGURE 50.—Placer deposits in the Ophir area.

1. Dodge Creek: Brooks (1916a, p. 65).
2-3. Ophir Creek: Mertie (1936, p. 176–179).
4. Anvil Creek: Mertie (1936, p. 190–191).
5. Victor Creek (Gulch): Mertie (1936, p. 191).
6. Spruce Creek: Mertie (1936, p. 179–181).
7. Tamarack Creek: Mertie (1936, p. 180).
8. Little Creek: Mertie (1936, p. 181–184).
9. Little Creek: Mertie (1936, p. 181–184), Joesting (1942, p. 42). No. 6 Pup: Joesting (1942, p. 39, 42).
10. Little Creek: Mertie (1936, p. 181–184).

11-12. Ganes Creek: Mertie (1936, p. 174–175, 184–188).
13. Spaulding Creek: Smith (1941b, p. 46). Spaulding Gulch: Smith (1937, p. 41–42).
14. Glacier Gulch: Maddren (1910, p. 70).
15. Last Chance Gulch: Maddren (1910, p. 70).
16. Mackie Creek: Mertie (1936, p. 188).
17-18. Yankee Creek: Mertie (1936, p. 188–190).
19. Bedrock Gulch: Smith (1937, p. 41–42).
20. Ester Creek: Mertie (1936, p. 192).

is possible that the country rock of the Ophir district is widely and diffusely mineralized, both by small gold-bearing quartz veins and in mineralized zones where little or no quartz is present." Modern stream gravels were the source of some of the gold that was mined, but the bulk of it came from deposits on bedrock benches, many of which have no surface expression, parallel to and near the streams. Some of the benches are not distinct and merge gently with bedrock surfaces beneath present stream channels. As much of the gold was in cracks along cleavage planes, the top foot or two of bedrock had to be taken up and processed with the basal part of the overlying gravel. Early mining was by drifting where the gravel was frozen, by small-scale hand methods, and by scraper plants, some of which were capable of moving large volumes of gravel. In 1921, the first dredge in the district was installed on Yankee Creek (17–18, fig. 50). Dredges were used on Ganes and Little (8–10, fig. 50) Creeks for many years. A dredge on Ganes Creek was still active in 1965. Some of the nonfloat operations employed as many men as dredges and may well have been as productive. Very few data are available on the composition of concentrates from streams in the Ophir area. Cinnabar has been reported from Anvil (4, fig. 50) and Victor (5, fig. 50) Creeks and much scheelite has been reported from Little Creek and a small tributary, No. 6 Pup (9, fig. 50).

The second most productive area in the Innoko district comprises streams that drain the northern Cripple Creek Mountains about 30 miles north-northeast of Ophir. The Cripple Creek Mountains are made up of sandstone, shale, greenstone, and chert intruded by a complex monzonitic pluton. Little is known about the distribution of rock types or the source of the gold in the placers. By analogy with geologically similar neighboring areas, it is probable that the gold was derived from mineralized quartz veinlets and small veins in the pluton and surrounding contact zone. There is a quartz-stibnite-cinnabar vein a short distance east of the Colorado Creek placers (7, fig. 37). From the distribution of mined areas visible on aerial photographs, it appears that both creek and bench placers were mined on Colorado and Cripple (6–7, fig. 37) Creeks and that a large mined area between Cripple and Bear (6, fig. 37) Creeks was probably a bench deposit. Several miles of ditches brought water to mining areas. No dredges were used in this part of the Innoko district, but nonfloat operations were extensive. The last reported activity was in 1967. There are no available data on the mineral composition of concentrates from any of these streams; the only heavy mineral mentioned in published reports is gold.

A third part of the Innoko district where there has been placer mining is (3–5, fig. 37) about 20 miles northwest of Ophir in an area drained by eastern tributaries of Tolstoi Creek (which in turn flows into the Dishna River) north of Mount Hurst. Most of the data on this area, at times in the past called the Tolstoi district, are based on an early reconnaissance by Harrington (1919a) made only a year and a half after the original discovery of gold on Boob Creek (3, fig. 37). The area is underlain by Mesozoic and Paleozoic limestone, chert, and volcanic and clastic rocks that were invaded by Tertiary mafic to monzonitic intrusive bodies. In addition to gold, platinum, amounting to about 1 percent of the precious-metal content of the concentrates, was recovered from Boob Creek. Other heavy minerals included magnetite, pyrite, garnet, and scarce cassiterite and cinnabar. Obsidian was found in the gravel. Most mining elsewhere in the area was on Madison and Esperanto Creeks (5, fig. 37). Since World War II, the only activity has been small nonfloat operations on these creeks and reports of what probably was mainly prospecting on one or two other streams. By 1965, even this had ceased.

Colors of gold have been found in other parts of the Innoko district, including the headwaters of the Dishna River (Maddren, 1911, p. 237), but successful placer mines were not developed.

KAIYUH DISTRICT

The Kaiyuh district (pl. 1, fig. 37) is the area drained by streams flowing west and north into the Yukon River from Shageluk Slough to (but excluding) the Yuko (Yuki) River and by streams flowing south into the Innoko River between Holikachuk and Dishkakat.

The district includes most of the Kaiyuh Hills and extensive low-lying areas between them and the Yukon River. The Kaiyuh Hills form an isolated range of northeast-trending mountains, generally from 1,000 to 2,250 feet in elevation with several summits that are a few hundred feet higher. About two-thirds of the district is made up of low swampy areas with many lakes, sloughs of the Yukon and Innoko Rivers, and sluggish streams following tortuous courses. Until the advent of the helicopter, access to most of the district was almost impossible except after freezup.

The Kaiyuh Hills are made up mainly of Precambrian(?) or lower Paleozoic(?) schist, quartzite, and recrystallized limestone and younger upper Paleozoic and (or) lower Mesozoic metamorphosed mafic extrusive and intrusive rocks (W. W. Patton, Jr., (oral commun., Oct. 6, 1970) with some interbedded chert, limestone, and graywacke. Cretaceous clastic rocks similar to those in the Koyukuk sedimentary basin to the north and west underlie low hills east of the lower Khotol River and an isolated peak just south of the Yukon River between Nulato and Galena. Several granitic plutons of Mesozoic or Tertiary age intruded the older rocks (Cass, 1959b, e; Maddren, 1910; Mertie, 1937a). The highlands, most of which are below tree line, are generally covered by frost-rived rubble, colluvium, and silt, some of which is probably loess; the lowlands are covered with thick alluvium. The district was not glaciated. Except for part of the Innoko Valley, where permafrost is discontinuous, the area is in a zone generally underlain by moderately thick to thin frozen ground.

Lode deposits have been found in two places in the Kaiyuh district (Berg and Cobb, 1967, p. 228–229, fig. 29). In the northern part of the district, argentiferous galena veins in schist were the source of 100–200 tons of silver-lead ore. In the southern part of the district, a molybdenite-bearing quartz vein was discovered in rhyolite porphyry, but no ore was mined.

The only placer mining in the Kaiyuh district was at Camp Creek (9, fig. 37) at the site of the abandoned village of Tlatskokot, where a small amount of gold was recovered immediately after World War II. The deposit must have been disappointing, as no activity was reported in later years. Farther south a Geological Survey party panned fine colors of gold from a tributary of the Kluklaklatna (Little Mud) River (8, fig. 37) near a granitic pluton.

KANTISHNA DISTRICT

The Kantishna district (pl. 1, fig. 39) is the area drained by the Kantishna River and small unnamed southern tributaries of the Tanana River between the Kantishna and the Zitziana Rivers.

The southern boundary of the district follows the crest of the Alaska Range and includes Mount McKinley, the highest peak in North America (20,300 ft). To the north, the mountains drop off abruptly to a gently northward-sloping lake-dotted area generally less than 2,000 feet above sea level. The Kantishna Hills, a group of rugged mountains less than 5,000 feet in elevation, extend about 20 miles north of the range.

North of the Denali fault zone, which is about 6 miles north and northwest of the crest of the Alaska Range and is hidden beneath the Muldrow Glacier for 17 miles, the geologic structure and rock types are extensions of those characteristic of the adjacent Bonnifield district. In the Kantishna district, however, the distribution of Tertiary rocks is much more restricted, and most of the Mesozoic felsic intrusive bodies are much larger than in the Bonnifield district. The rocks south of the Denali fault zone are generally unmetamorphosed Mesozoic sedimentary rocks that were intruded by a large granitic batholith near Mount McKinley and by possibly related stocks (Reed, 1961). The lowlands are covered by Quaternary unconsolidated glaciofluvial and eolian deposits. The Alaska Range and part of the piedmont were covered by ice during several

Pleistocene glaciations, but the lowlands were ice free. Today most of the high country is glacier covered, and spectacular ice tongues, of which the Muldrow Glacier is the longest, extend down to elevations of less than 4,000 feet. The mountains and the Kantishna Hills are in a zone of discontinuous permafrost. The lowlands are largely underlain by moderately thick to thin continuous permafrost; in areas of coarse-grained deposits, only isolated masses are perennially frozen.

Lodes in the Kantishna district have yielded small amounts of gold, silver, antimony, and lead ore (Berg and Cobb, 1967, p. 229–231, fig. 31). Most of the lodes are quartz veins in schistose rocks and carry gold, argentiferous galena, and other sulfide minerals; one group of deposits about 25 miles east of Kantishna, consists of argentiferous galena and other sulfides in contact-metamorphosed limestone. Stibnite ore was mined from deposits at Stampede (near locality 6, fig. 39) and on Slate Creek a few miles southwest of Kantishna. Cinnabar is a major constituent of one lode high in the mountains near the head of Birch Creek.

Placer gold was found in the Kantishna district in 1903 somewhere on Chitsia Creek, a tributary that enters the Kantishna River a few miles northwest of the abandoned settlement of Nineteenmile (Reed, 1961, p. A26). Early in the summer of 1905, gold in paying quantities was discovered on Glacier (6, fig. 51) and neighboring creeks, setting off a stampede of several thousand people during which practically all of the ground in the Kantishna Hills was staked. By the fall of 1906, most of the stampeders had left the country; only about 50 men, those who had staked the best claims, were left. The most productive years were 1906 (3,600 fine ounces of gold) and 1940 (nearly 4,000 ounces). The 1940 production was mainly by two dragline operations on Caribou Creek (9, 10, fig. 51). The total production from the Kantishna district through 1960 was probably between 45,000 and 50,000 fine ounces. Since 1960, there has been small-scale mining on Crooked (5, fig. 39), Eureka (4, fig. 51), and Friday, (5, fig. 51) Creeks; the amount of gold recovered probably was small. No activity was reported in 1968.

Except for Crooked Creek and two neighboring creeks, all the placer mining in the district was on streams that drain the area near Kantishna where lode deposits were mined or prospected. The heavy minerals in the placer concentrates undoubtedly came directly from nearby sources, for all the metallic minerals in the placers have been found in lodes in the basins of the creeks that were mined; in fact, several lodes were uncovered during placer

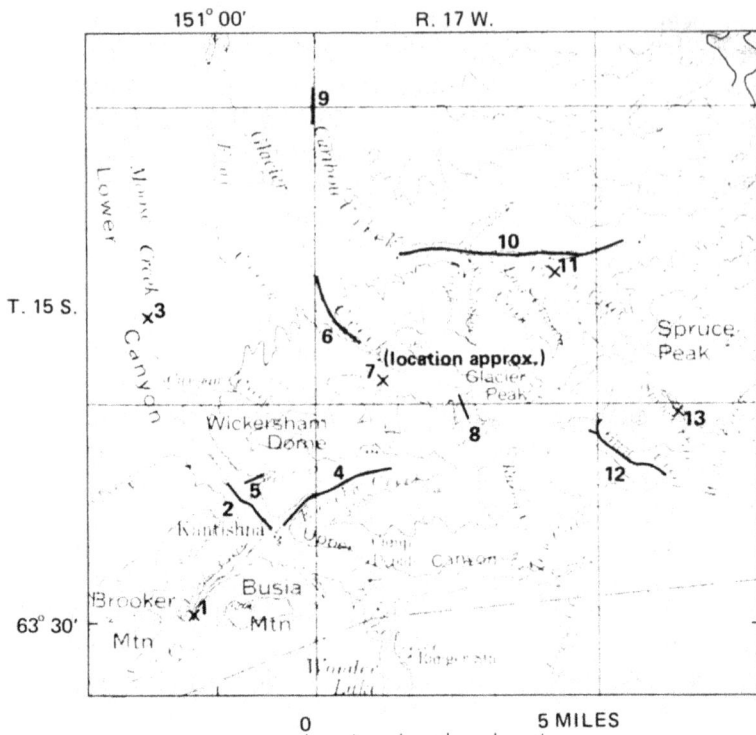

FIGURE 51.—Placer deposits in the Kantishna area.

1. Eldorado Creek: Capps (1919a, p. 88).
2. Moose Creek: Capps (1919a, p. 88–89), Brooks and Capps (1924, p. 41).
3. Moose Creek: Prindle (1907, p. 217).
4. Eureka Creek: Capps (1919a, p. 85–87).
5. Friday Creek: Brooks (1916c, p. 42), Capps (1919a, p. 87–88).
6. Glacier Creek: Capps (1919a, p. 90–92), Wells (1933, p. 371).
7. Twenty-two Gulch (Pup): Wells (1933, p. 371), Smith (1942b, p. 49).
8. Yellow Creek: Wells (1933, p. 371).
9-10. Caribou Creek: Prindle (1907, p. 218–219), Capps (1919a, p. 92–93), Smith (1942b, p. 49).
11. Crevice Creek: Prindle (1907, p. 218).
12. Glen Creek: Capps (1919a, p. 83–85), Davis (1923, p. 116).
13. Spruce Creek: Prindle (1907, p. 214–215).

operations. Most of the gravel mined was in the beds of the streams, although bench gravels on Glacier Creek are auriferous. Scheelite was a common constituent of concentrates from Little Moose and Stampede Creeks (6, fig. 39). Small nuggets of native silver were recovered from Little Moose Creek.

KOYUKUK DISTRICT

The Koyukuk district (pl. 1, fig. 52) is the area drained by the Koyukuk River and its tributaries above and including the Kanuti River.

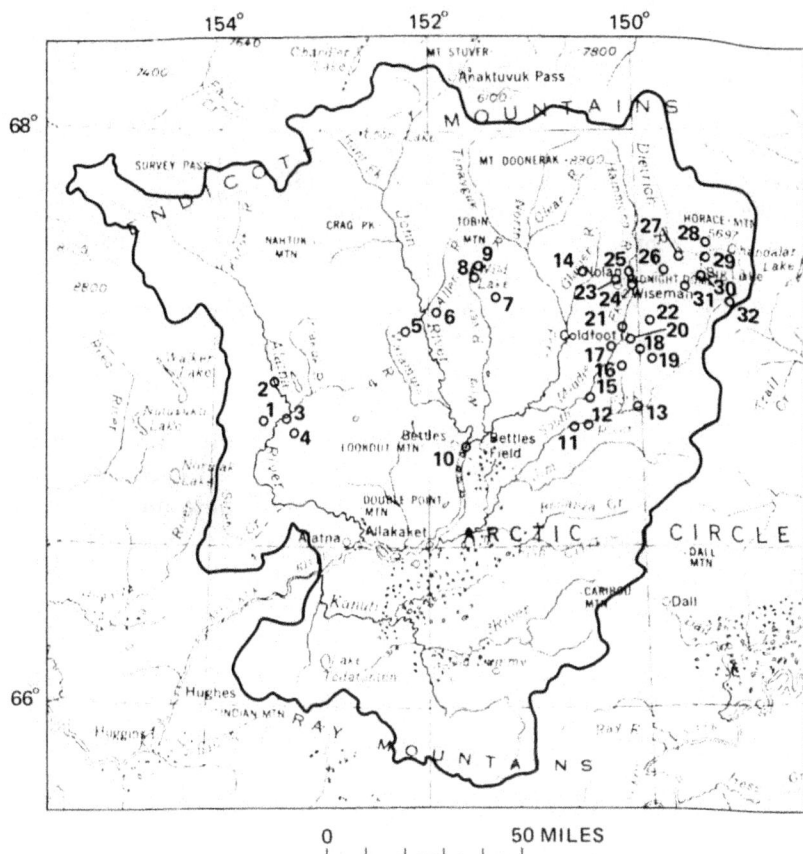

FIGURE 52.—Placer deposits in the Koyukuk district.

1. Helpmejack Creek: Mendenhall (1902, p. 50).
2-3. Alatna River: Smith (1913, p. 143), Smith and Mertie (1930, p. 334-335).
4. Rockybottom Creek: Mendenhall (1902, p. 50).
5. Midas Creek: Maddren (1913, p. 110), Alaska Department of Mines (1952, p. 61).
6. Crevice Creek: Maddren (1913, p. 110), Brosgé and Reiser (1960).
7. Jay Creek: Brooks (1916a, p. 65). Rye Creek: White (1952b, p. 8-11).
8. Lake Creek: Maddren (1913, p. 109), Smith and Mertie (1930, p. 333), Joesting (1943, p. 17).
9. Spring Creek: Maddren (1913, p. 110), Brosgé and Reiser (1960). Surprise Creek: Smith (1933a, p. 39-40).
10. Bettles Bar: Reed (1938, p. 163-164).
11. Gold Bench: Maddren (1913, p. 105-107), Reed (1938, p. 153-154), White (1952b, p. 11), Wedow and others (1953, p.

3), Nelson, West, and Matzko (1954, p. 18-19). Ironside Bench (Bar): Maddren (1913, p. 107), Reed (1938, p. 154-155).
12. Davis Creek: Maddren (1913, p. 70, 107). Rock Creek: Reed (1938, p. 160-161).
13. Eagle Cliff: Maddren (1913, p. 107). Grubstake Bar: Maddren (1913, p. 107).
14. Mascot Creek: Maddren (1913, p. 108-109).
15. Chapman Creek: Maddren (1913, p. 85). Mailbox Creek: Smith (1937, p. 45). Tramway Bar: Maddren (1913, p. 84-85), Brosgé and Reiser (1960).
16. Rose (Rosie) Creek: Maddren (1913, p. 86).
17. Porcupine Creek: Maddren (1913, p. 86), Brosgé and Reiser (1960).
18. Myrtle Creek: Maddren (1913, p. 86-89).
19. Slate Creek: Maddren (1913, p. 86-88), Brosgé and Reiser (1964).

The Koyukuk district, one of the largest in the Yukon River region, extends from the Brooks Range on the north southward to the Ray Mountains. Between these highlands are lower uplands and broad, poorly drained lowlands, of which the Kanuti Flats in the southern part of the district is the largest. The crest of the Brooks Range, which reaches elevations of 6,000 to more than 7,000 feet, is broken by several passes, the lowest of which, Anaktuvuk Pass at the head of the John River, is below 2,200 feet. The Kanuti Flats and similar terrain near the confluence of the South Fork of the Koyukuk with the main river are generally less than 600 feet above sea level. In the Ray Mountains, summit elevations are generally between 3,000 and 4,000 feet.

Geologically, the northern two-thirds of the district is a westward extension of the Chandalar district; bedded rocks of Paleozoic age underly much of the area and a belt of metamorphic rocks, mainly of greenscist facies and of Devonian age, extends across the southern foothills of the Brooks Range (Brosgé and Reiser, 1960, 1964; Herreid, 1969; Patton and Miller, 1966, 1970). Except near Big Lake and west of the head of Malamute River (W. P. Brosgé, oral commun., Aug. 27, 1970), this part of the Koyukuk district was not invaded by large granitic plutons, although dikes are fairly common in some places. In the Ray Mountains and the highlands at the head of the Dall River, however, large Cretaceous quartz monzonite plutons and smaller bodies of hypabyssal rocks intruded older metasedimentary rocks. The Koyukuk sedimentary basin is underlain mainly by Cretaceous marine graywacke and

---

FIGURE 52.—Continued.

20. Clara Gulch: Maddren (1913, p. 89), Smith (1936, p. 43).
21. Emma Creek: Maddren (1913, p. 69, 90-91). Kelly Gulch: Maddren (1913, p. 89).
22. Marion Creek: Maddren (1913, p. 90).
23. Archibald Gulch: Maddren (1913, p. 92-93). Fay Gulch: Maddren (1913, p. 92-94). Nolan Creek: Maddren (1913, p. 92-94). Quartz Pup: Alaska Department of Mines (1946, p. 38). Smith Creek: Maddren (1913, p. 92-94), Joesting (1943, p. 17). Thompson Gulch: Alaska Department of Mines (1958, p. 48). Webster Gulch: Alaska Department of Mines (1946, p. 44). Wiseman Creek: Maddren (1913, p. 91-92).
24. Confederate Gulch: Maddren (1913, p. 95). Union Gulch: Maddren (1913, p. 64, 95).
25. Buckeye and Goldbottom Gulches: Maddren (1913, p. 96-97). Hammond River:

Maddren (1913, p. 95-96). Swift Gulch (Creek): Maddren (1913, p. 69, 96-97). Vermont Creek: Maddren (1913, p. 69, 97-98), Brosgé and Reiser (1960).
26. Gold Creek: Schrader (1904, p. 105), Maddren (1913, p. 99-102). Linda Creek: Maddren (1913, p. 102-104). Sheep Creek (Gulch): Maddren (1913, p. 99), Brosgé and Reiser (1964).
27. Emory Creek: Maddren (1913, p. 104-105).
28. Mule Creek: Maddren (1913, p. 104-105).
29. Eightmile Creek: Maddren (1913, p. 105), Joesting (1943, p. 18). Garnet Creek: Maddren (1913, p. 69, 104).
30. Jim Gulch (Pup): Maddren (1913, p. 70, 108). Lake Creek: Brosgé and Reiser (1964). Wakeup Creek: Smith (1939b, p. 55).
31. Boer (Bore) Creek (Gulch): Maddren (1913, p. 108).
32. Sawlog Creek: Brosgé and Reiser (1964).

mudstone and nonmarine clastic and tuffaceous beds. Pleistocene ice from the Brooks Range and from cirque glaciers in the Ray Mountains covered much of the higher parts of the district; the lowlands are floored by Pleistocene glaciofluvial and Holocene flood-plain deposits. Modern glaciers are restricted to cirques in the highest parts of the Brooks Range. Most of the area is in zones of continuous permafrost except for some of the highlands where permafrost is continuous. Unfrozen zones in several valleys at about the latitude of Wiseman caused considerable difficulty to drift miners.

Lodes containing gold and silver and antimony, copper, lead, and manganese minerals have been discovered in the part of the district underlain by metamorphic rocks. The lodes are mainly quartz and quartz-stibnite veins in phyllite and schist and galena in crystalline limestone (Berg and Cobb, 1967, p. 234, fig. 35). The only production was a few tons of stibnite ground sluiced from the surface of a vein (Joesting, 1943, p. 17) near Wiseman. Geochemical investigations (Brosgé and Reiser, 1970a, b; Patton and Miller, 1970) resulted in the discovery of anomalous concentrations of several metals in stream-sediment, soil, and bedrock samples from the northeastern and southern parts of the district. Bedrock occurrences of sulfide minerals and chrome spinel were also found.

Placer gold was found on bars of the Koyukuk River and some of its forks between 1885 and 1890, but the first major discoveries were on Myrtle Creek (18, fig. 52) and Hammond River (25, fig. 52) in 1899 and 1900, respectively. Since then, there has been mining in every year. Production through 1961, the last year for which any figure is available, was probably between 270,000 and 295,000 fine ounces of gold, of which more than 85 percent came from streams in a small area north of, and within 10 miles of, Wiseman. Other places where there was more or less successful placer mining are near Wild Lake, on tributaries of the John River near the mouth of the Allen River, on tributaries of the Middle Fork of the Koyukuk near Coldfoot and Big Lake, and on bars of the Middle and South Forks of the Koyukuk River. Gold on the Alatna River and its tributaries Helpmejack and Rocky-bottom Creeks (1–4, fig. 52) could not be mined profitably.

In the area north of Wiseman, the most productive streams were Nolan Creek and its left-limit tributaries (23, fig. 52) and Hammond River and its tributary Vermont Creek (25, fig. 52). The heavy minerals in the placers were derived from nearby auriferous quartz and quartz-stibnite veins, many of which were uncovered during mining operations. Normal stream placers developed in the headwater reaches of the main creeks and in their tributary gulches, but later damming of the larger creeks by ice tongues that, during several advances, followed the Dietrich River and Middle Fork valleys and backed up into the valley of Wiseman Creek caused many complications in the lower reaches. Placer deposits in Nolan Creek, for example, occupy a V-shaped gulch that was later filled and buried by sediments graded to a higher temporary base level. Buried bedrock channels of streams tributary to the Middle Fork are truncated, suggesting that the main valley was glacially scoured, cutting off the mouths of tributary channels before they were reburied. Any gold in these truncated channels was removed; if present, it may have been a source of the gold on bars farther down the Middle Fork. Although there are fewer data on placers in other parts of the district, some may have had similar histories. Glaciation caused other derangements of preglacial drainage. For example, the placer on Linda Creek (26, fig. 52) probably represents the lower part of an old channel of Gold Creek (26, fig. 52) that was later encountered by Linda Creek, a small consequent stream occupying part of an interglacial or postglacial cutoff of the Middle Fork. Bench deposits on bedrock terraces have been mined on Nolan Creek and several other streams. These deposits were laid down by streams graded to glacial deposits that choked the main valleys after ice fronts had withdrawn northward. Some of the bars on the lower Middle and South Forks of the Koyukuk River formed when the streams flowed at a higher level. Gold Bench (11, fig. 52) is about 30 feet above the present grade of the South Fork and consists of gravel on false bedrock. In the early 1900's, about 100 acres was mined and about 6,000 ounces of gold was recovered by hand methods using water brought from a small tributary of the South Fork. Although Maddren (1913, p. 106–107) considered the most likely source of the gold to have been hills to the south, it seems reasonable that some gold, at least, was reconcentrated from glacial deposits. A similar deposit at Tramway Bar (15, fig. 52) on the Middle Fork rests on nonauriferous conglomerate and sandstone 80–100 feet above river level. It was being mined as recently as 1965. Gold credited to small creeks such as Buckeye Gulch (25, fig. 52) actually came from bench deposits along the larger streams into which they flow.

Large-scale mining has been rare in the Koyukuk district; use of heavy equipment other than bulldozers was uncommon in recent years. Deep frozen gravel was mined from drifts, many of which were driven using wood fires that caused at least one operator to die from lack of oxygen. Shallow ground was mined mainly with

hand tools, though there was a little hydraulicking and considerable groundsluicing. The largest operation in the district, on
Myrtle Creek, used a dragline and bulldozers and was active as
recently as 1953. By 1969, however, only five men were reported to
be mining in the entire district—two men each were working on
Linda and Porcupine (17, fig. 52) Creeks and one was working on
Vermont Creek.

Placer concentrates from the Koyukuk district have not been
studied as intensively as those from many other parts of Alaska.
Stibnite is a common constituent of concentrates from the creeks
near Wiseman; it is so common on Smith Creek (23, fig. 52), a
tributary of Nolan Creek, that 5 tons sluiced from creek gravels
and from a lode exposed in the creek bottom was shipped during
World War II. Small native silver nuggets and pieces of native
copper (probably derived from nearby greenstone sills) weighing
as much as 7 pounds were found on Mule Creek (28, fig. 52). Native
copper, native bismuth, and scheelite were reported from Lake
Creek (8, fig. 52). Concentrate samples from Gold Bench, studied
more thoroughly than those from anywhere else in the district,
contained trace amounts of pyrite, chalcopyrite, cinnabar, rutile,
cassiterite, scheelite, monazite, uranothorianite, and various nonmetallic accessory minerals. Monazite, andalusite, kyanite, and
scheelite, as well as sulfide and rock minerals, were identified in
concentrates from Rye Creek (7, fig. 52). Large gold nuggets are
more common than in most parts of Alaska; some nuggets worth
several hundred dollars have been reported from many creeks in
the district.

Placer-gold mining has been reported from many more localities,
particularly near the South Fork of the Koyukuk River, than are
shown on the map (fig. 52). Some occurrences could not be plotted
because descriptions are ambiguous or because creek names were
changed several times and old names cannot be correlated with
modern ones. It is unlikely, however, that a significant amount of
gold was produced from these localities, either singly or as a group.

MARSHALL DISTRICT

The Marshall district (pl. 1, fig. 37) comprises the area drained
by the Yukon River and its tributaries below Paimiut Slough and
by streams flowing into the Bering Sea between Hazen and St.
Michael Bays and a small area northwest of the Yukon River between (but excluding) the Koserefsky River and the lower end of
Paimiut Slough.

The district includes most of the delta of the Yukon River, a low
marshy area studded with lakes and traversed by sluggish streams.

Small but rugged mountain masses protrude through the alluvial cover and reach elevations greater than 2,000 feet in the Askinuk Mountains near Cape Romanzof and the Kuzilvak Mountains about 80 miles west of Marshall. The part of the district north of the Yukon River consists of a series of northeast-trending ridges with crest elevations generally between 1,000 and 2,000 feet that are separated by valleys having long, straight segments.

The following summary of the regional geology of the Marshall district is based on recent geologic reconnaissance maps by Coonrad (1957), Hoare and Condon (1966, 1968, 1971a, b), and Hoare and Coonrad (1959b).

Mesozoic, mainly Cretaceous, marine and nonmarine clastic rocks and andesitic volcanic and volcaniclastic rocks underlie most of the district. Paleozoic and Mesozoic volcanic and sedimentary rocks are exposed in the hills north of the Yukon River east of Marshall. Younger Cretaceous felsic bodies, including a granodiorite batholith in the Askinuk Mountains, intruded the bedded rocks and thermally metamorphosed them near some contacts. The youngest consolidated rocks in the district, mainly on Stuart Island and near St. Michael, are horizontal Quaternary basalt flows and associated fragmental rocks and their source cones, some of which are practically unmodified by erosion. Quaternary alluvial, deltaic, and marine sand and silt make up most of the Yukon Delta and floor most river valleys and a large, low swampy area north of Marshall near Kuyukutuk. Glacial deposits in the Askinuk Mountains record the presence of small Pleistocene cirque glaciers; the rest of the district was ice free. The entire area is in a zone generally underlain by moderately thick to thin permafrost. A well near Marshall penetrated 125 feet of perennially frozen ground.

Lodes in the Marshall district contain gold and lead, molybdenum, copper, and tungsten minerals, but none has been productive (Berg and Cobb, 1967, p. 235). The most thoroughly explored deposit, the Arnold lode, about 7 miles east of Marshall, is in altered volcanic rock and consists of mineralized quartz-calcite veins as much as a foot thick. This and similar lodes in the same area were the probable sources of the gold in nearby placers.

Prospectors who visited parts of the Marshall district during the early 1900's probably found colors of gold in many places (Harrington, 1918, p. 56), but minable placer deposits were not discovered until 1913, when gold was found first on Wilson Creek (11, fig. 37) and then on neighboring creeks. Actual mining began in 1914 and continued until at least as recently as 1965. All the productive placer operations in the district were on Wilson Creek and its tributaries (11, fig. 37), Willow Creek (10, fig. 37) and tributaries of Kako Creek (12, fig. 37), where gold was discovered in about 1920. Total production from the Marshall district cannot be stated, because it usually was combined with that from the Anvik district. Through 1961, the last year for which data are available, gold recovered from streams in both districts was probably between 115,000 and 120,000 fine ounces; of this, possibly 80 percent was from the Marshall district. The principal operations were near Marshall on Willow Creek and on Bobtail and Montezuma Creeks (12, fig. 37), tributaries of Kako Creek. Most mining was by nonfloat methods, though drift and hand methods were also employed. In addition to gold, an unknown but undoubtedly small amount of platinum, probably derived from the greenstone country rock, was recovered from Willow (10, fig. 37) and Wilson and Disappointment (11, fig. 37) Creeks. Other heavy minerals included magnetite, hematite, ilmenite, and, from Elephant Creek (11, fig. 37), scheelite. Cinnabar was an uncommon constituent of the concentrates from Bobtail Creek. No source of the cinnabar has been found, but it probably is a lode similar to the one about 30 miles to the north on Wolf Creek Mountain in the Anvik district.

MELOZITNA DISTRICT

The Melozitna district (pl. 1, fig. 47) is the area drained by northern tributaries of the Yukon River between and including the Melozitna and Ray Rivers.

The Melozitna district consists mainly of rolling ridges generally 2,500–4,000 feet in elevation above which a few isolated groups of mountains rise as much as 1,500 feet. The upper parts of some of the major rivers flow in wide swampy valleys choked with alluvium and commonly less than 1,000 feet in elevation. The streams reach the Yukon River through narrow canyons.

Bedrock in the greater part of the district is metamorphosed sedimentary and volcanic rocks, mainly of probable Paleozoic age (Cass, 1959c, d; Eakin, 1916), similar to those that underlie most of the upland between the Yukon and Tanana Rivers. In the eastern part of the district, a large area northwest of the Yukon River is underlain by the principally volcanic Rampart Group of probable Permian age (Brosgé, Lanphere, and others, 1969). Much of the valley of the Melozitna River is in Cretaceous marine and nonmarine clastic rocks of the southeastern margin of the Koyukuk sedimentary basin. Granitic plutons of Mesozoic age intruded and further metamorphosed the older rocks in many parts of the district. Pleistocene cirque glaciers developed in the higher parts of the Kokrines Hills and Ray Mountains but have now disappeared. The district is in zones of moderately thick to thin continuous or discontinuous permafrost. Many of the placer deposits, however, are in shallow gravels that are not permanently frozen.

Lode deposits in the Melozitna district include a gold-quartz vein at Gold Hill, near Grant Creek (16, fig. 47), and galena veins about 25 miles up the Yukon River from Tanana and near Tozimoran Creek (17, fig. 47) (Berg and Cobb, 1967, p. 236, fig. 34).

Placer gold was discovered on streams in the Melozitna district in 1907, but mining was sporadic and production data are incomplete. The output for many years was combined with that from the Rampart or Hot Springs districts or was buried in totals for other districts with only one or two producers. Gold production through 1961 was probably about 5,000 fine ounces, but this figure may be in error by a few thousand ounces. After 1961, a small nonfloat operation was reported annually until 1964; since that time the district appears to have been nonproductive. Cassiterite is a common constituent of many of the placers, and from time to time some tin concentrates were saved, but it is not certain that any were marketed.

All the placer deposits are within a few miles of known or postulated granitic intrusive bodies, but no lode sources for either the gold or the cassiterite in the placers have been found despite diligent search by prospectors and Federal geologists and engineers (Chapman and others, 1963; Thomas and Wright, 1948a, b). Most of the tin investigations were on Morelock Creek and its tributaries (18, fig. 47) and on Tozimoran (17, fig. 47) and Mason (12, fig. 47) Creeks.

Gold has been mined from thin bench deposits and shallow stream gravels of these creeks and from other localities shown in figure 47; the most recent activity was on Golden (14, fig. 47) and Grant Creeks (16, fig. 47). All the mining was small scale. Other streams, mainly ones that flow into the Yukon River, were prospected and colors of gold were reported, but many of the reports were unsubstantiated, and the locations of the possibly auriferous deposits were only generally described.

Concentrates from creeks in the Melozitna district have not been studied in detail except to determine cassiterite content, so very few data on their constituents are available. Magnetite, ilmenite, hematite, and garnet, in addition to gold and cassiterite, were identified in samples from the basin of Morelock Creek. Tourmaline accompanies cassiterite in veinlets in micaceous quartzite pebbles in alluvium on Tozimoran Creek.

RAMPART DISTRICT

The Rampart district (pl. 1, fig. 47) is the area drained by southern tributaries of the Yukon River between and including Fish and Hamlin Creeks.

The district is characterized by broad ridges 1,500–3,000 feet in elevation separated by wide valleys. The western part of the area is more rugged than the eastern, having deeper, narrower valleys and several isolated peaks rising to elevations greater than 3,500 feet.

Much of the Rampart district is underlain by Paleozoic clastic rocks, chert, and limestone (Eakin, 1913a; Mertie, 1934, 1937b; Prindle, 1908a; Prindle and Hess, 1906). The probably Permian Rampart Group (Brosgé, Lanphere, and others, 1969), which is mainly volcanic but includes some interbedded limestone, chert, and clastic rocks, is exposed along the Yukon River and underlies about half of the district. Cretaceous clastic rocks extend from the adjoining Hot Springs district into the headwater basins of Minook and Troublesome Creeks, and Tertiary terrestrial deposits occupy a local basin near the Yukon River near Rampart. Granitic plutons of Tertiary age were emplaced along the southern boundary of the district west of Sawtooth Mountain, and felsic and small mafic dikes, some possibly associated with the volcanic rocks of the Rampart Group, are fairly common in the Minook Creek area. From east to west the pre-Tertiary bedded rocks of the district are progressively more metamorphosed; some are now garnet schists.

Much of the district is covered by Quaternary residual deposits, alluvium, and loess. Extensive stream terraces, particularly in the Minook Creek area, are covered with gravel and finer alluvial material. This part of the Yukon-Tanana Upland was not glaciated and is in a zone of discontinuous permafrost. Most of the bench gravels are frozen, whereas the deposits in streambeds generally are not.

No lode mines have been developed in the Rampart district, although antimony and manganese minerals have been found in bedrock and in float thought to be close to its source (Berg and Cobb, 1967, p. 236–237, fig. 34; Thomas, 1965). Burand and Saunders (1966), on the basis of stream-sediment sampling near Minook Creek, found indications of three parallel mineralized zones distinguished by anomalous amounts of lead, copper, and copper in association with zinc.

Placer gold was first found in the Rampart district in 1882 by the Schieffelin brothers, the original discoverers of gold at Tombstone, Ariz., on stream bars said to yield $10 per man per day. The locations of these discoveries are not known, but probably they were bars of Minook and Hess Creeks and their tributaries. In 1893, John Minook found coarse gold in Minook and Hess Creeks and other nearby streams, and another prospector did a little mining in a gulch (which one is not known) near the head of Hess Creek. Sustained mining did not begin for another 3 years, but has since been carried on in nearly every year. Despite the long history of the district, total production through 1961, the last year for which data have been made public, was only about 80,000–90,000 fine ounces of gold, most of which was recovered before 1920. No mining was reported in 1968, but in the preceding year there were five one- or two-man operations. Except for small amounts of gold from Willow Creek (22, fig. 47) and other localities vaguely described as being in Hess Creek basin, all of the production was from streams in the basins of Minook and Troublesome Creeks.

Minook Creek is deeply incised in a narrow valley. On the east valley wall, there are four prominent terraces and several others that are less conspicuous. Near the mouth of the creek, the terraces are from 10 to about 1,000 feet above stream level. Some can be traced up tributary valleys and all approach the grade of Minook Creek upstream. Pliocene(?) gravel as much as 100 feet thick caps the highest terrace. Parts of the gravel on this terrace, preserved between eastern tributaries of Minook Creek, are called, in order upstream, Yukon Bar, Idaho Bar, California Bar, and McDonald or Florida Bar. All contain auriferous gravel and contributed much of the gold to stream and lower bench placers, but only on Idaho Bar (9, fig. 53) was there any mining of the higher gravels. The ultimate sources of the gold and most of the other heavy minerals in the placers were probably quartz veins (some of which yield a few colors of gold when crushed and panned) and altered mafic rock (greenstone). Placer deposits on Minook Creek are generally of too low grade to be mined by small-scale methods and were worked at only a few places (3, fig. 53). Several plans for the installation of a dredge or large-scale hydraulic plant were reported, but none reached an operational stage. Large hydraulic mines were never developed on tributaries of Minook Creek because water is usually scarce in their upper reaches. Probably more than half of the gold produced in the district came from Little Minook Creek (8, fig. 53), where barren material was ground-sluiced off and the gold-bearing gravel and top 18 inches or so of fractured bedrock were shovelled into sluice boxes. Other placers mined include stream gravel on other creeks in the area and bench

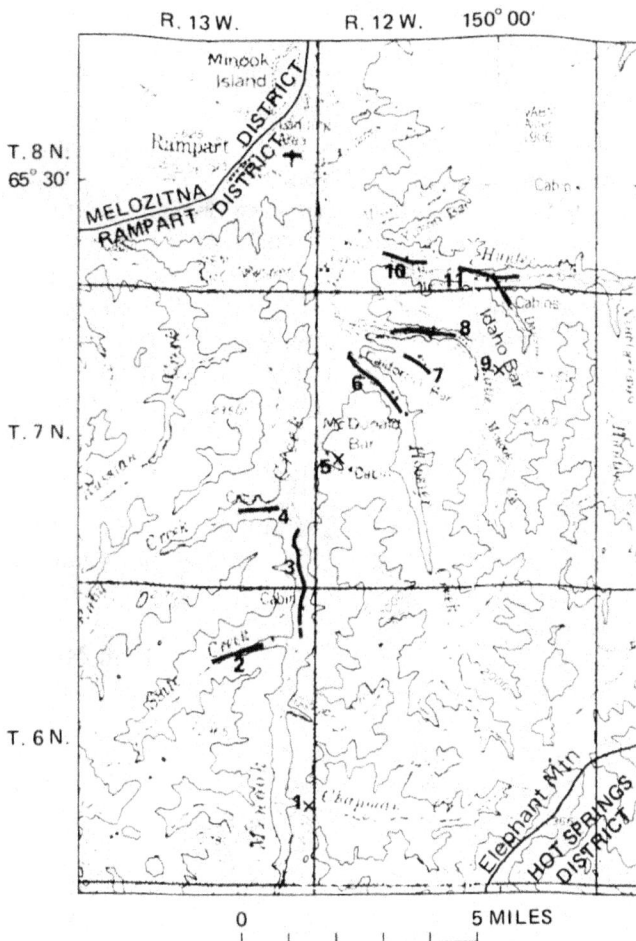

FIGURE 53.—Placer deposits in the Minook Creek area.

1. Chapman Creek: Ellsworth and Davenport (1913, p. 222), Burand and
   Saunders (1966, p. 5).
2. Slate Creek: Hess (1908, p. 65, 81-82),
   Mertie (1934, p. 165, 187-188).
3. Minook Creek: Mertie (1934, p. 165, 174-
   177).
4. Ruby Creek: Mertie (1934, p. 188-189),
   Burand and Saunders (1966, p. 5).
5. Florida Creek: Mertie (1934, p. 175, 191).
6. Hoosier Creek: Mertie (1934, p 165, 186-
   187), Waters (1934, p. 234).

7. Little Minook Junior Creek: Mertie (1934,
   p. 165, 175, 184-185), Waters (1934, p.
   234).
8. Little Minook Creek: Mertie (1934, p. 165,
   181-183), Waters (1934, p. 232-234).
9. Idaho Bar: Mertie (1934, p. 183-184),
   Waters (1934, p 234-235).
10. Hunter Creek: Mertie (1934, p. 165, 177-
    181), Waters (1934, p. 232).
11. Dawson Creek: Mertie (1934, p. 180).
    Hunter Creek: Mertie (1934, p. 165, 177-
    181), Waters (1934, p. 232).

gravel along Hunter (10, 11, fig. 53) and other creeks tributary to
Minook Creek and on Quail Creek (19, fig. 47).

The concentrates from most of the placer deposits in the western Rampart district contain many heavy minerals in addition to

gold. Barite and garnet are present in most, and iron oxide minerals, chrome spinel, and pyrite are common. Galena was reported from Hunter, Little Minook, Little Minook Junior (7, fig. 53), and Troublesome (21, fig. 47) Creeks, and argenite (Ag_2S) was reported from Little Minook Creek. Native copper nuggets are common in the concentrates of Little Minook Creek and were identified in samples from other deposits. Miners found native silver nuggets (one weighing 2 ounces) in sluice boxes on Ruby Creek (4, fig. 53) and smaller nuggets elsewhere in the area. Small amounts of scheelite, chromite, cinnabar, and native bismuth and tetradymite (Bi_2Te_2S) have been reported but are not common. Cassiterite is a major component of a sample from Quail Creek and a little was identified in samples from Hunter and Troublesome (21, fig. 47) Creeks.

RUBY DISTRICT

The Ruby district (pl. 1, fig. 47) comprises the Poorman area and the area drained by southern tributaries of the Yukon River from (and including) the Yuko (Yuki) River on the west to Kallands on the east.

The district includes the eastern part of the Kaiyuh Mountains, the northern part of the Kuskokwim Mountains, and a low, poorly drained area immediately south of the Yukon River. The mountains are generally less than 2,000 feet in elevation except in the southern part of the district near the head of the Sulukna River where they are more rugged and are about 1,000 feet higher. A few isolated summits in the northern part of the area are above 3,000 feet and Von Frank Mountain on the divide between the Yukon and Kuskokwim drainages, is 4,508 feet in elevation. The lowland is a poorly drained area of low relief, mainly less than 500 feet above sea level, through which streams meander in wide belts characterized by many oxbow lakes.

This summary of the geology of the Ruby district is based on maps and reports by Cass (1959c–1959e), Eakin (1918), Mertie (1937a), and Mertie and Harrington (1924). The district is mainly underlain by Precambrian(?) and Paleozoic limestone, schist, and other metamorphic rocks and by altered volcanic rocks. Cretaceous clastic rocks exposed north of the Yukon River west of Ruby may underlie some of the northern part of the district. Near Poorman, Cretaceous chert and argillite are interbedded with coarser clastic and volcanic rocks. Cretaceous and possibly Tertiary granitic plutons invaded the older rocks in many parts of the district, but only small intrusive bodies have been found in areas with known mineral deposits. The district was not glaciated.

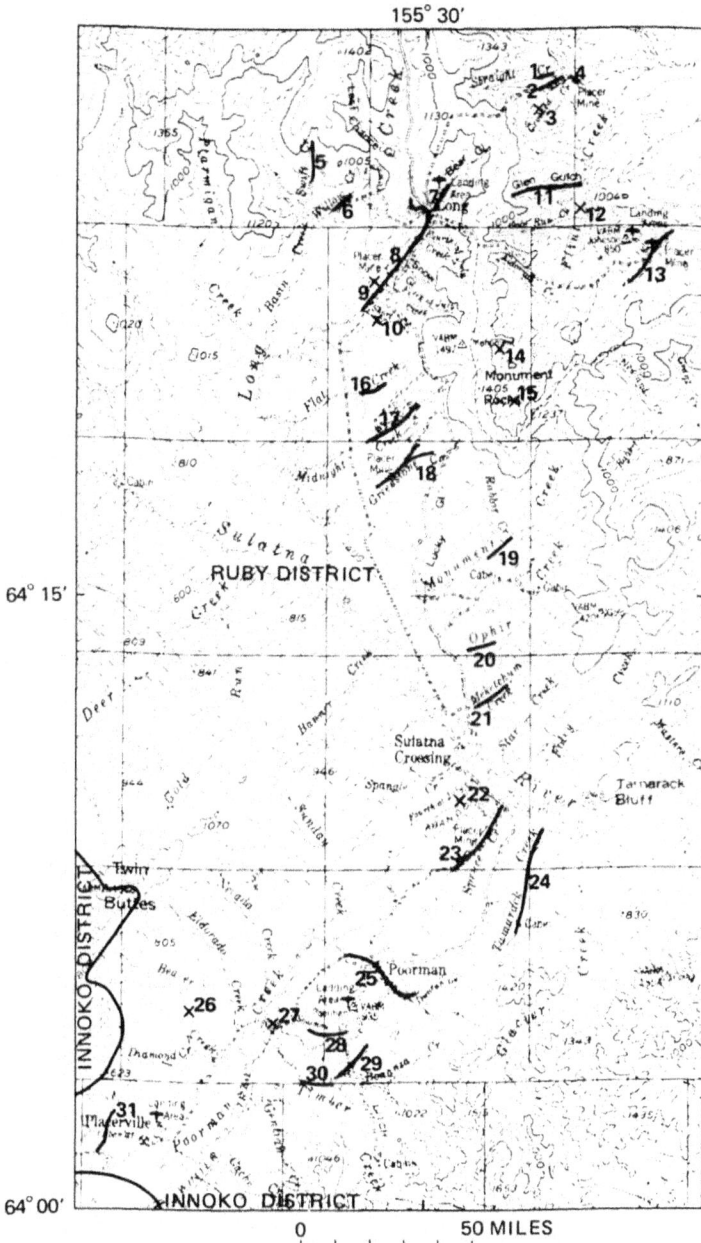

FIGURE 54.—Placer deposits in the Long-Poorman area.

1. Straight Creek: Chapman, Coats, and Payne (1963, p. 42–44).
2. Crooked Creek: Chapman, Coats, and Payne (1963, p. 42–43).
3. Lucky Creek: Eakin (1914b, p. 367).

4. Birch Creek: White and Stevens (1953, p. 3–4, 7), Chapman, Coats, and Payne (1963, p. 37, 40, 42–44).
5. Swift Creek: Mertie (1936, p. 145, 155).

Large areas are covered by weathered residual material and unconsolidated alluvium and terrace deposits, most of which are perennially frozen.

Even though the area was not glaciated, the geomorphic history of the Ruby district is complex and is not fully understood. A long period of erosion produced a mature surface with a thick cover of residual material containing large amounts of vein quartz fragments and pieces of chert and other resistant rocks. Where streams flowed on bedrock, natural riffles concentrated heavy minerals that had been carried to the streams by mass wasting. After a change in regimen, the streams deposited a thick fill of sediments that buried much of the preexisting surface. Because most of this fill remains, the modern creeks flow on bedrock only in their headwater reaches and in a few places where they were randomly superposed across buried spurs. As the present streams do not lie above the deepest parts of their old valleys except by random coincidence, most prospecting has been hit-or-miss.

The only described lode in the district is a silver-lead deposit in metamorphic rocks near the head of Beaver Creek about 10 miles south of Ruby (Brown, 1926b). Other mineralized zones are

---

· FIGURE 54.—Continued

6. Willow Creek: Mertie and Harrington (1916, p. 243).
7. Bear Gulch (Pup): Mertie (1936, p. 146, 148-151).
8. Long Creek: Mertie (1936, p. 146-151), Mertie (1937a, p. 173).
9. Fifth of July Creek: Chapman, Coats, and Payne (1963, p. 49).
10. Short Creek: Mertie (1936, p. 151), Chapman, Coats, and Payne (1963, p. 48-49).
11. Glen Gulch: Mertie (1936, p. 158).
12. Flint Creek: Mertie (1936, p. 148-149, 157-158), White and Stevens (1953, p. 4-9), Chapman, Coats, and Payne (1963, p. 40, 49-50).
13. Trail Creek: Mertie (1936, p. 148-149, 156-157), Joesting (1943, p. 20), Chapman, Coats, and Payne (1963, p. 37, 50).
14. Granite Creek: Moffit (1927, p. 33).
15. Flint Creek: Mertie (1936, p. 148-149, 157-158), White and Stevens (1953, p. 4-9), Chapman, Coats, and Payne (1963, p. 40, 49-50).
16. Flat Creek: Chapman, Coats, and Payne (1963, p. 48).
17. Midnight Creek: Joesting (1943, p. 20), Chapman, Coats, and Payne (1963, p. 37, 40-42).
18. Greenstone Creek: Mertie (1936, p. 146, 152-153), Chapman, Coats, and Payne

(1963, p. 37, 40, 47-48). Greenstone Gulch: Mertie (1936, p. 152-153).
19. Monument Creek: Mertie (1936, p. 153-154), Chapman, Coats, and Payne (1963, p. 51).
20. Ophir Creek: Eakin (1914b, p. 367-368), Cass (1959c).
21. Meketchum Creek: Mertie (1936, p. 155).
22. Fourth of July Creek: Brooks (1918, p. 57).
23. Spruce Creek: Mertie and Harrington (1916, p. 244), Chapin (1919).
24. Tamarack Creek: Mertie and Harrington (1916, p. 244-245), Chapin (1919). Willow Gulch: Brooks (1915, p. 58).
25. Duncan Creek: Mertie (1936, p. 163). Little Pup: Mertie (1936, p. 163). Poorman Creek: Mertie and Harrington (1916, p. 247-248, 265), Mertie (1936, p. 158-162, 165-166). Tenderfoot Creek: Eakin (1914b, p. 368-369).
26. Beaver Creek: Mertie (1936, p. 169).
27. Poorman Creek: Mertie and Harrington (1916, p. 247-248, 265), Mertie (1936, p. 158-162, 165-166).
28. Solomon Creek: Mertie (1936, p. 164-165), White and Stevens (1953, p. 1, 9).
29. Flat Creek: Chapin (1919), Mertie (1936, p. 166-167).
30. Timber Creek: Smith (1941b, p. 50).
31. Moose Creek: Mertie (1936, p. 168-169).

probably hidden by the thick mantle of residual debris which generally supports a heavy growth of moss and effectively hides bedrock, even on most ridges.

Placer gold was discovered on Ruby Creek (23, fig. 47) in 1907, on Long Creek 3 years later (8, fig. 54), on Poorman Creek in 1912 (25, fig. 54), and on Moose Creek (31, fig. 54) in 1920, although the pay streak was not located there until 1931. Total gold production from the Ruby district through 1960 was about 390,000 fine ounces (1.9 percent of Alaska's total placer-gold output). In 1961, the only more recent year for which data are available, two mines produced a total of 1,439 ounces of gold and a little alloyed silver. Annual production since then, mainly from nonfloat operations on Long and Greenstone (18, fig. 54) Creeks, probably has been comparable. About 10 short tons of cassiterite concentrate was recovered from gold placers, mainly during World Wars I and II, but the amount actually sold is not known. A little byproduct platinum (probably only a few ounces) came from Granite Creek (14, fig. 54).

The only dredge in the district operated in 1916 and 1917 on Greenstone Creek and its major headwater fork, Greenstone Gulch (18, fig. 54), where the auriferous gravel was shallow. Most of the placers, however, were too deep for common types of surface mining and were worked from shafts and drifts. An old channel of Long Creek was extensively mined, as were similar deposits near Poorman and on Moose Creek. Shallow gravels in the headwater parts of tributaries of Long Creek and the Sulatna River could be worked with scrapers or bulldozers, or by hand methods. The shortage of water with much natural head precluded extensive hydraulic mining. Water for sluicing had to be pumped and reused in many operations except during short periods in the spring when snowmelt was plentiful.

Heavy minerals in placer concentrates include cassiterite, which has been found in samples from most creeks tributary to the Sulatna River and in two creeks near Poorman. Platinum occurs in Granite Creek, scheelite in Trail (13, fig. 54) and Midnight (17, fig. 54) Creeks, and allanite in Birch (4, fig. 54) and Flint (12, fig. 54) Creeks. Native bismuth was identified in concentrates from Birch and Glacier (24, fig. 47) Creeks. Quartz and iron-oxide minerals made up a large part of the concentrates from many creeks; sulfide minerals generally were not common.

Gold has been reported from many creeks in the southern and eastern parts of the district. Most of these reports, however, are either unconfirmed or stated in such general terms that the occurrences cannot be located within many miles. Any production from them was certainly very small.

SHEENJEK DISTRICT

The Sheenjek district (pl. 1, fig. 38) is bounded on the west by the drainage divide east of the East Fork of the Chandalar River, on the north by the crest of the Brooks Range, and on the east by the Alaska-Yukon boundary. Its southern border is an arbitrary line extending eastward from Venetie on the Chandalar River to Graphite Point on the Porcupine River, beyond which it follows the divide between the Porcupine and Black Rivers to the Canadian border.

The peaks marking the crest of the Brooks Range are 4,000–7,000 feet in elevation except near the Alaska-Yukon boundary, where they are lower and are separated by a few passes below 3,000 feet. Southward, the Brooks Range merges with rounded ridges near the Porcupine River and low (less than 1,000 feet) swampy flats in the southwestern part of the district.

The following resumé of the geology of the Sheenjek district is based on reconnaissance mapping and topical studies by Brabb (1969), Brosgé, Dutro, and others (1960), Brosgé and Reiser (1962, 1965, 1968, 1969), Brosgé, Reiser, and others (1966), and Reiser and Tailleur (1969).

The bedded rocks in the district are mainly Paleozoic and Mesozoic clastic, carbonate, and volcanic rocks; some of the sedimentary rocks may be as old as Precambrian and some of the volcanic rocks may be as young as Quaternary. Many of the bedded rocks in the southern part of the district have been regionally metamorphosed to schist and phyllite. A large granitic pluton that has been radiometrically dated as being 295–335 million years old (Carboniferous) intruded lower Paleozoic rocks near the head of the Rapid River. A similar, but smaller pluton is east of the upper Coleen River. A large mafic igneous complex of Jurassic age (155–168 million years old on the basis of potassium-argon analyses) intruded Paleozoic rocks in the west-central part of the district. The lowlands consist of Quaternary alluvial fan and related terrace deposits; aeromagnetic data suggest that large areas are underlain by mafic igneous rocks (Brosgé, Brabb, and King, 1970).

Only the northern part of the district was glaciated, although glaciofluvial deposits make up much of the Quaternary deposits in the lowlands. Some of the peaks at the head of the Sheenjek River still support small glaciers. Most of the district is in zones of continuous permafrost, though in parts of the northern Yukon Flats, perennially frozen ground may be locally absent.

No mineral resources have been developed in the Sheenjek district. During a geochemical reconnaissance of the Coleen and Table Mountain quadrangles, Brosgé and Reiser (1968) found anomalous concentrations of lead, copper, zinc, molybdenum, and tin in stream-sediment and crushed bedrock samples. They found galena at Bear Mountain (about 25 miles west of Ammerman Mountain) and chalcopyrite a few miles south of the Brooks Range divide between the extreme headwater forks of the Coleen River (W. P. Brosgé, oral commun., Aug. 18, 1970). Mertie (1930, p. 138) reported that gold had been found on bars of the Coleen River, but he did not specify which bars. Pleistocene deposits on Pass Creek (5, fig. 38) contain a little gold, but not in minable concentrations. In 1948, White (1952a) conducted a brief reconnaissance for radioactive deposits during which he found traces of uraniferous minerals, one of which was tentatively identified as clarkeite, in stream-concentrate and bedrock samples from a tributary of the Rapid River (6, fig. 38) and from Sunaghun Creek (7, fig. 38). These localities are close to the south boundary of the large granitic pluton near the head of the Rapid River.

TOK DISTRICT

The Tok district (pl. 1, fig. 41) is the area drained by southern tributaries of the Tanana River between (and including) Berry Creek and the Tetlin and Kalutna Rivers. Most of the district is on the northeast flank of the Alaska Range, which slopes from Mount Kimball (10,300 ft) to lowlands along the Tanana River that are only about 1,500 feet above sea level. Ice, remnants of Pleistocene glaciers that covered most of the Alaska Range, persists around Mount Kimball and other high mountains and flows several miles down some of the larger valleys. Discontinuous permafrost underlies the district.

Bedrock in the Tok district is mainly Paleozoic and Mesozoic clastic and volcanic rocks, large Mesozoic granitic plutons, and smaller bodies of older, more mafic intrusive rocks. The lowlands are mantled by unconsolidated fluvial and glacial deposits largely derived from the Alaska Range. Two lode deposits (Berg and Cobb, 1967, p. 238–239, fig. 32) were found in the early 1900's, but they were never mined. One was a stibnite prospect in a shear zone in schist on a tributary of the Tok River, and the other was an occurrence of gold, silver, copper, and possibly nickel near the highway between Gulkana and Tok Junction.

Although most of the Tok district probably has been prospected, only two occurrences of placer gold have been reported; neither was productive. Shallow gravel at the summit of Mentasta Pass (39, fig. 41) contained a little gold, and colors were found on Moose Creek (40, fig. 41).

TOLOVANA DISTRICT

The Tolovana district (pl. 1, fig. 44) is the area drained by southwest-flowing tributaries of the Tanana River from Dugan Creek on the west to the Tolovana River on the east. The drainage basin of the Chatanika River above the mouth of the Tatalina River is not included. The district is bounded on the north by an arbitrary line dividing the lowlands of the Yukon Flats from higher ground drained mainly by upper Beaver Creek.

Most of the district is characterized by broad, even-topped ridges with an average summit elevation of about 2,000 feet, above which elongate mountain masses rise 2,000–3,000 feet. In general, the valley floors are wide and occupied by underfit streams. The lower part of the Tolovana River valley is a broad, low, lake-dotted marshy area through which the main stream and the lower parts of its tributaries flow in tortuous channels.

Most of the geologic data on the Tolovana district are in reports by R. L. Foster (1966, 1969) and Mertie (1937b). The southern part of the district, geologically an extension of the adjacent Fairbanks and Circle districts, is a metamorphic terrane of schists and related rocks of Precambrian(?) and early Paleozoic age. To the northwest, these rocks are succeeded by northeastward-trending belts of slightly metamorphosed Paleozoic clastic, volcanic, and carbonate rocks and chert. Lower Cretaceous clastic rocks underlie an area near Sawtooth Mountain in the western part of the district. Many granitic and dioritic plutons and dikes, some probably of Mesozoic or Tertiary age, are in the district. At least one pluton, that at Amy Dome near Livengood, is older than late Middle Devonian and could not have been a source of mineralization in the exposed surrounding rocks, as all of them are younger than the intrusive body. A belt of serpentinite bodies of Devonian(?) age extends northeastward from Livengood for many miles. Evidence of more than one Pleistocene glaciation has been found on the higher peaks of the White Mountains, on Mount Schwatka, and on Sawtooth Mountain, but all are now ice free. The district is mainly within a zone of discontinuous permafrost, but many south-facing slopes are not perennially frozen. The flats in the lower valley of the Tolovana River generally are underlain by moderately thick to thin permafrost and numerous isolated masses of frozen ground.

Lodes in the Tolovana district have been found only in the area immediately around Livengood and on Sawtooth Mountain (Berg and Cobb, 1967, p. 239–240, fig. 33; Foster and Chapman, 1967). They contain gold, silver, antimony, mercury, chromium, nickel, and iron, but the only ore recovered was small amounts of antimony ore from Livengood and Sawtooth Mountain and a little cinnabar from near the head of Olive Creek (10, fig. 55). A serpentinite body between the head of the Tolovana River and Beaver Creek contains anomalously large amounts of nickel and detectable platinum and palladium (Foster, R. L., 1969). Many of the placer cuts near Livengood have uncovered quartz and calcite veins that contain gold and sulfide minerals, and similar veins have been found in nearby hills.

Although placer gold had long been reported from upper tributaries of Beaver Creek, the first substantial discovery was not made until 1910, when minable deposits were found on Ophir Creek (12, fig. 44). During the ensuing small stampede, all the ground in the Nome Creek (12–14, fig. 44) and neighboring basins was staked. In 1914, workable gold placers were discovered on Livengood Creek (2, fig. 55) and some of its tributaries; thus, the Tolovana is one of the most recently developed major placer districts in Alaska. Total gold production reported through 1960 was about 380,000 fine ounces, about 1.9 percent of the total for the State. The actual production of the Tolovana district was certainly higher than that reported, possibly by many thousands of ounces, as an indeterminate amount of the output of Nome Creek had been credited to the Fairbanks district.

Nome Creek and its major tributary Ophir Creek are similar to neighboring creeks in the Fairbanks district. The gold in Nome Creek probably had a common source with that in Sourdough Creek in the Fairbanks district, as both drain an area where a small granitic pluton intruded Precambrian(?) schists. Both stream and bench deposits are auriferous. Dredges, the first of which was installed in 1926, were operated on Nome Creek for many years. A sample of dredge concentrates contained abundant cassiterite and topaz and sparse monazite and tourmaline. From time to time there were rumors of gold on other streams in the basin of Beaver Creek, but no valuable placers were located.

The most important feature in the erosional and depositional Quaternary history of the Livengood area (fig. 55) was the development of a mature surface that was largely buried by later

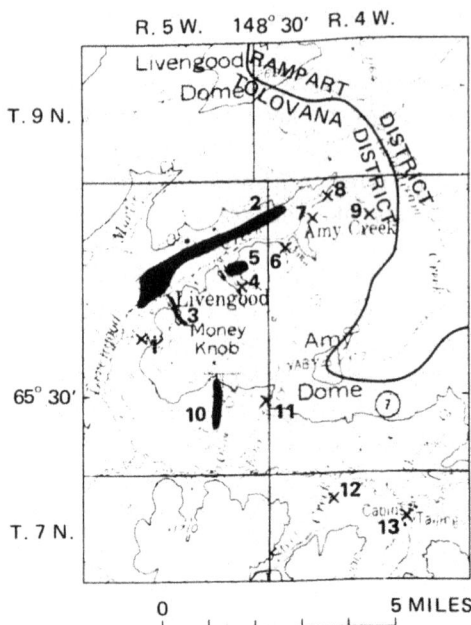

FIGURE 55.—Placer deposits in the Livengood area.

1. Lillian Creek: Mertie (1918a, p. 270-271), Overbeck (1920, p. 181-183), Joesting (1942, p. 14, 26, 39).
2. Franklin Creek: Brooks (1916b, p. 208). Livengood Creek: Mertie (1918a, p. 259-268), Overbeck (1920, p. 178-181, 183), Joesting (1942, p. 14, 17, 34, 39). Myrtle Creek: Brooks (1918, p. 56).
3. Ruth Creek: Brooks (1918, p. 22), Mertie (1918a, p. 269-271, 273-274), Overbeck (1920, p. 183), Joesting (1942, p. 17, 26, 39), Wedow, Killeen, and others (1954, p. 11).
4. Glen Gulch: Mertie (1918a, p. 269).
5. Gertrude Creek: Mertie (1918a, p. 209).
6. Lucille Creek: Smith (1936, p. 40), Joesting (1942, p. 17).
7. Amy Creek: Overbeck (1920, p. 181, 184), Joesting (1942, p. 14, 17).
8-9. Goodluck (Lucky) Creek: Smith (1941b, p. 49), Joesting (1942, p. 17), Wedow, White, and others (1954, p. 2-3).
10. Olive Creek: Mertie (1918a, p. 271-272, 274), Overbeck (1920, p. 182-183), Joesting (1942, p. 17, 26, 39).
11. Ester Creek (Lucky Gulch): Mertie (1918a, p. 271-272), Smith (1932, p. 36).
12. Steel Creek: Joesting (1943, p. 20).
13. Wilbur Creek: Smith (1942b, p. 46).

sediments. The divide between Livengood and Hess Creeks shifted back and forth in response to successive stream captures. The richest placers were gravels representing old stream-channel material on what are now buried bedrock benches that the present streams have not completely exhumed. Some deposits in the beds of modern streams and residual placers near the head of a few creeks that drain Money Knob and Amy Dome have been mined.

The longest and richest old channel is on an extensive bench northwest of Livengood Creek (2, fig. 55) that has been traced in drift mines, drill holes, and surface workings from opposite the mouth of Amy Creek to Livengood Creek below the mouth of

Myrtle Creek. Near the settlement of Livengood, this channel was extensively dredged for several years and is still (1970) being worked on a small scale. Water was brought across the divide at the head of the stream from a tributary of Hess Creek. Bench deposits were mined on Olive Creek (10, fig. 55) and on several northwest-flowing tributaries of Livengood Creek. Gold reported to have been mined from Myrtle Creek (2, fig. 55) and other right-limit tributaries of Livengood Creek probably was reconcentrated from the Livengood bench deposit. Of the streams that have supported profitable mining, only Steel and Wilbur Creeks (12, 13, fig. 51) do not drain Money Knob or Amy Dome. Data on these creeks are scarce and the source of the gold in them is not known.

Many heavy minerals have been identified in concentrates from creeks in the Livengood area. They include magnetite, hematite, ilmenite, and limonite, abundant chromite and chrome spinel, and cinnabar, stibnite, and other sulfide minerals. Less common are scheelite and cassiterite, one or both of which have been found in Lillian (1, fig. 55), Ruth (3, fig. 55), Livengood, Steel, and Goodluck (8–9, fig. 55) Creeks, monazite found in Livengood Creek, and a niobium-titanium-uranium-rare earth mineral found in Goodluck Creek.

Mining in the Tolovana district has become progressively less active as old timers have died or moved away. In 1968, four-man nonfloat operations were reported on Amy and Livengood Creeks and a few men were prospecting.

YUKON FLATS DISTRICT

The Yukon Flats district (pl. 1, fig. 38) comprises the relatively flat basin of the Yukon River from Circle to the site of Fort Hamlin (a few miles downstream from Stevens Village), the drainage basin of the Porcupine River below Graphite Point, and the entire drainage basins of the Dall, Hodzana, and Hadweenzic Rivers.

Most of the Yukon Flats district is a swampy basin, generally less than 1,000 feet in elevation. The basin fill is mainly late Tertiary and Quaternary glaciofluvial material from the Brooks Range to the north, stream deposits derived from the upper Yukon River basin, and reworked material forming loess blankets and sand dunes (Williams, 1962). Recent aeromagnetic investigations (Brosgé, Brabb, and King, 1970) suggest that much of the area beneath the surficial deposits is underlain by mafic igneous rocks. The northwestern part of the district is a rolling upland composed of schist, quartzite, and crystalline limestone of early

Paleozoic or possibly Precambrian age that were cut by granitic and dioritic plutons and dikes (Cobb, 1969a, p. 4; Williams, 1962, p. 297–302). Sedimentary and volcanic rocks and associated mafic intrusive rocks exposed along the Yukon River in the western part of the district were once thought to be Mississippian in age, but recently have been shown to be probably Permian (Brosgé, Lanphere, and others, 1969); some of the intrusive rocks are probably as young as Triassic. Tertiary continental rocks were deposited in local basins near the Dall and Hodzana Rivers.

The only lode mineralization reported from the Yukon Flats district is on Trout Creek (8, fig. 38), from which specimens containing quartz, pyrite, sphalerite, and molybdenite were sent to the Geological Survey by a prospector in 1923 (Smith, 1942a, p. 197).

The only placer mining reported in the district was a small operation on Slate Creek and its tributary Trout Creek (8, fig. 38) near the divide at the head of the Pitka Fork of the Hodzana River. All that is known about this mining is that a few men worked in each of several years during the 1950's and 1960's. Production, which must have been small, was probably credited to the Chandalar district.

REFERENCES CITED

Alaska Department of Mines, 1940, Report of the Commissioner of Mines, biennium ended December 31, 1940: Alaska Dept. Mines, 92 p.
——— 1946, Report of the Commissioner of Mines, biennium ended December 31, 1946: Alaska Dept. Mines, 50 p.
——— 1948, Report of the Commissioner of Mines, biennium ended December 31, 1948: Alaska Dept. Mines, 50 p.
——— 1950, Report of the Commissioner of Mines, biennium ended December 31, 1950: Alaska Dept. Mines, 57 p.
——— 1952, Report of the Commissioner of Mines, biennium ended December 31, 1952: Alaska Dept Mines, 65 p.
——— 1954, Report of the Commissioner of Mines, biennium ended December 31, 1954: Alaska Dept. Mines, 110 p.
——— 1956, Report of the Commissioner of Mines, biennium ended December 31, 1956: Alaska Dept. Mines, 103 p.
——— 1958, Report of the Commissioner of Mines, biennium ended December 31, 1958: Alaska Dept. Mines, 83 p.
Alaska Division of Mines and Geology, 1968, Report for the year 1968: Alaska Div. Mines and Geology, 67 p.
Alaska Division of Mines and Minerals, 1960, Report for the year 1960: Alaska Div. Mines and Minerals Rept. 1960, 88 p.
——— 1962, Report for the year 1962: Alaska Div. Mines and Minerals Rept. 1962, 119 p.
——— 1964, Report for the year 1964: Alaska Div. Mines and Minerals Rept. 1964, 107 p.

——— 1966, Report for the year 1966: Alaska Div. Mines and Minerals Rept. 1966, 115 p.

——— 1967, Report for the year 1967: Alaska Div. Mines and Minerals Rept. 1967, 98 p.

Anderson, Eskil, 1945, Asbestos and jade occurrences in the Kobuk River region, Alaska: Alaska Dept. Mines Pamph. 3–R, 26 p.

——— 1947, Mineral occurrences other than gold deposits in northwestern Alaska: Alaska Dept. Mines Pamph. 5–R, 48 p.

Asher, R. R., 1969, Geology and geochemistry of part of the Iron Creek area, Solomon D–6 quadrangle, Seward Peninsula, Alaska: Alaska Div. Mines and Geology Geochem. Rept. 18, 19 p.

Atwood, W. W., 1911, Geology and mineral resources of parts of the Alaska Peninsula: U.S. Geol. Survey Bull. 467, 137 p.

Barnes, F. F., 1966, Geology and coal resources of the Beluga-Yentna region, Alaska: U.S. Geol. Survey Bull. 1202–C, p. C1–C54.

Barth, T. F. W., 1956, Geology and petrology of the Pribilof Islands, Alaska: U.S. Geol. Survey Bull. 1028–F, p. 101–160.

Barton, W. R., 1962, Columbium and tantalum, a materials survey: U.S. Bur. Mines Inf. Circ. 8120, 110 p.

Bates, R. G., and Wedow, Helmuth, Jr., 1953, Preliminary summary report of thorium-bearing mineral occurrences in Alsaka: U.S. Geol. Survey Circ. 202, 13 p.

Berg, H. C., and Cobb, E. H., 1967, Metalliferous lode deposits of Alaska: U.S. Geol. Survey Bull. 1246, 254 p.

Berg, H. C., Eberlein, G. D., and MacKevett, E. M., Jr., 1964, Metallic mineral resources, in Mineral and water resources of Alaska: U.S. 88th Cong., 2d sess., Senate Comm. Interior and Insular Affairs, Comm. Print, p. 95–125.

Berryhill, R. V., 1963, Reconnaissance of beach sands, Bristol Bay, Alaska: U.S. Bur. Mines Rept. Inv. 6214, 48 p.

Blake, W. P., 1868, Geographical notes upon Russian America and the Stickeen River: U.S. 40th Cong., 2d sess., H. Ex. Doc. 177, pt. 2, 19 p.

Brabb, E. E., 1969, compiler, Preliminary geologic map of the Black River quadrangle, east-central Alaska: U.S. Geol. Survey open-file map.

Brabb, E. E., and Churkin, Michael, Jr., 1965, Preliminary geologic map of the Eagle D–1 quadrangle, Alaska: U.S. Geol. Survey open-file map.

——— 1969, Geologic map of the Charley River quadrangle, east-central Alaska: U.S. Geol. Survey Misc. Geol. Inv. Map I–573, scale 1:250,000.

Brabb, E. E., and Miller, D. J., 1962, Reconnaissance traverse across the eastern Chugach Mountains, Alaska: U.S. Geol. Survey Misc. Geol. Inv. Map I–341, scale 1:96,000.

Brew, D. A., Loney, R. A., and Muffler, L. J. P., 1966, Tectonic history of southeastern Alaska: Canadian Inst. Mining and Metallurgy Spec. V. 8, p. 149–170.

Brooks, A. H., 1901, A reconnaissance of the Cape Nome and adjacent gold fields of Seward Peninsula, Alaska, in 1900: U.S. Geol. Survey Spec. Pub., p. 1–180.

——— 1907a, The mining industry in 1906: U.S. Geol. Survey Bull. 314, p. 19–39.

——— 1907b, The Kougarok region: U.S. Geol. Survey Bull. 314, p. 164–181.

——— 1907c, The Circle precinct, Alaska: U.S. Geol. Survey Bull. 314, p. 187–204.

—— 1911, The Mount McKinley region, Alaska, with descriptions of the igneous rocks and of the Bonnifield and Kantishna districts, by L. M. Prindle: U.S. Geol. Survey Prof. Paper 70, 234 p.

—— 1912, The mining industry in 1911: U.S. Geol. Survey Bull. 520, p. 17–44.

—— 1913, The mining industry in 1912: U.S. Geol. Survey Bull. 542, p. 18–51.

—— 1914, The Alaskan mining industry in 1913: U.S. Geol. Survey Bull. 592, p. 45–74.

—— 1915, The Alaskan mining industry in 1914: U.S. Geol. Survey Bull. 622, p. 15–68.

—— 1916a, The Alaska mining industry in 1915: U.S. Geol. Survey Bull. 642, p. 16–71.

—— 1916b, Preliminary report on the Tolovana district: U.S. Geol. Survey Bull. 642, p. 201–209.

—— 1916c, Antimony deposits of Alaska: U.S. Geol. Survey Bull. 649, 67 p.

—— 1918, The Alaskan mining industry in 1916: U.S. Geol. Survey Bull. 662, p. 11–62.

—— 1922, The Alaskan mining industry in 1920: U.S. Geol. Survey Bull. 722, p. 7–67.

—— 1923, The Alaskan mining industry in 1921: U.S. Geol. Survey Bull. 739, p. 1–50.

—— 1925, Alaska's mineral resources and production, 1923: U.S. Geol. Survey Bull. 773, p. 1–52.

Brooks, A. H., and Capps, S. R., 1924, The Alaskan mining industry in 1922: U.S. Geol. Survey Bull. 755, p. 3–56.

Brosgé, W. P., Brabb, E. E., and King, E. R., 1970, Geologic interpretation of reconnaissance aeromagnetic survey of northeastern Alaska: U.S. Geol. Survey Bull. 1271–F, p. F1–F14.

Brosgé, W. P., Dutro, J. T., Jr., Mangus, M. D., and Reiser, H. N., 1960, Geologic map of the eastern Brooks Range, Alaska: U.S. Geol. Survey open-file map.

Brosgé, W. P., Lanphere, M. A., Reiser, H. N., and Chapman, R. M., 1969, Probably Permian age of the Rampart Group, central Alaska: U.S. Geol. Survey Bull. 1294–B, p. B1–B18.

Brosgé, W. P., and Reiser, H. N., 1960, Progress map of the geology of the Wiseman quadrangle, Alaska: U.S. Geol. Survey open-file map.

—— 1962, Preliminary geologic map of the Christian quadrangle, Alaska: U.S. Geol. Survey open-file map.

—— 1964, Geologic map and section of the Chandalar quadrangle, Alaska: U.S. Geol. Survey Misc. Geol. Inv. Map I–375, scale 1:250,000.

—— 1965, Preliminary geologic map of the Arctic quadrangle, Alaska: U.S. Geol. Survey open-file map.

—— 1968, Geochemical reconnaissance maps of granitic rocks, Coleen and Table Mountain quadrangles, Alaska: U.S. Geol. Survey open-file rept., 4 sheets.

—— 1969, Preliminary geologic map of the Coleen quadrangle, Alaska: U.S. Geol. Survey open-file map.

—— 1970a, compilers, Chemical analyses of rock and soil samples from the Chandalar and eastern Wiseman quadrangles, Alaska: U.S. Geol. Survey open-file rept., 8 p.

———— 1970b, compilers, Chemical analyses of stream sediment samples from the Chandalar and eastern Wiseman quadrangles, Alaska: U.S. Geol. Survey open-file rept., 5 p., 51 computer sheets.

Brosgé, W. P., Reiser, H. N., Dutro, J. T., Jr., and Churkin, Michael, Jr., 1966, Geologic map and stratigraphic sections, Porcupine River Canyon, Alaska: U.S. Geol. Survey open-file rept., 4 sheets.

Brosgé, W. P., Reiser, H. N., and Estlund, M. B., 1970, Chemical analyses of stream sediment samples from the Sadlerochit-Jago Rivers area, Mt. Michelson and Demarcation Point quadrangles, Alaska: U.S. Geol. Survey open-file rept., 44 p.

Brosgé, W. P., and Tailleur, I. L., 1970, Depositional history of northern Alaska [abs.]: Am. Assoc. Petroleum Geologists Bull., v. 54, no. 5, p. 837–838.

Brown, J. S., 1926a, The Nixon Fork country: U.S. Geol. Survey Bull. 783, p. 97–144.

———— 1926b, Silver-lead prospects near Ruby: U.S. Geol. Survey Bull. 783, p. 145–150.

Burand, W. M., 1965, A geochemical investigation between Chatanika and Circle Hot Springs, Alaska: Alaska Div. Mines and Minerals Geochem. Rept. 5, 11 p.

Burand, W. M., and Saunders, R. H., 1966, A geochemical investigation of Minook Creek, Rampart district, Alaska: Alaska Div. Mines and Minerals Geochem. Rept. 12, 18 p.

Burk, C. A., 1965, Geology of the Alaska Peninsula—island arc and continental margin: Geol. Soc. America Mem. 99, pt. 1, 250 p.; pt. 2, geol. map; pt. 3, tectonic map.

Byers, F. M., Jr., 1957, Tungsten deposits in the Fairbanks district, Alaska: U.S. Geol. Survey Bull. 1024–I, p. 179–216.

Cady, W. M., Wallace, R. E., Hoare, J. M., and Webber, E. J., 1955, The central Kuskokwim region, Alaska: U.S. Geol. Survey Prof. Paper 268, 132 p.

Cairnes, D. D., 1915, Upper White River district, Yukon: Canada Geol. Survey Mem. 50, 191 p.

Capps, S. R., 1912, The Bonnifield region, Alaska: U.S. Geol. Survey Bull. 501, 64 p.

———— 1913, The Yentna district, Alaska: U.S. Geol. Survey Bull. 534, 75 p.

———— 1915, The Willow Creek district, Alaska: U.S. Geol. Survey Bull. 607, 86 p.

———— 1916a, The Chisana-White River district, Alaska: U.S. Geol. Survey Bull. 630, 130 p.

———— 1916b, The Turnagain-Knik region: U.S. Geol. Survey Bull. 642, p. 147–194.

———— 1919a, The Kantishna region, Alaska: U.S. Geol. Survey Bull. 687, 116 p.

———— 1919b, Mineral resources of the western Talkeetna Mountains: U.S. Geol. Survey Bull. 692, p. 187–205.

———— 1925, An early Tertiary placer deposit in the Yentna district: U.S. Geol. Survey Bull. 773, p. 53–61.

———— 1934, Notes on the geology of the Alaska Peninsula and Aleutian Islands: U.S. Geol. Survey Bull. 857–D, p. 141–153.

———— 1935, The southern Alaska Range: U.S. Geol. Survey Bull. 862, 101 p.

—— 1937, Kodiak and adjacent islands, Alaska: U.S. Geol. Survey Bull. 880–C, p. 111–184.

—— 1940, Geology of the Alaska Railroad region: U.S. Geol. Survey Bull. 907, 201 p.

Cass, J. T., 1959a, Reconnaissance geologic map of the Norton Bay quadrangle, Alaska: U.S. Geol. Survey Misc. Geol. Inv. Map I–286, scale 1:250,000.

—— 1959b, Reconnaissance geologic map of the Unalakleet quadrangle, Alaska: U.S. Geol. Survey Misc. Geol. Inv. Map I–288, scale 1:250,000.

—— 1959c, Reconnaissance geologic map of the Ruby quadrangle, Alaska: U.S. Geol. Survey Misc. Geol. Inv. Map I–289, scale 1:250,000.

—— 1959d, Reconnaissance geologic map of the Melozitna quadrangle, Alaska: U.S. Geol. Survey Misc. Geol. Inv. Map I–290, scale 1:250,000.

—— 1959e, Reconnaissance geologic map of the Nulato quadrangle, Alaska: U.S. Geol. Survey Misc. Geol. Inv. Map I–291, scale 1:250,000.

Cathcart, S. H., 1920, Mining in northwestern Alaska: U.S. Geol. Survey Bull. 712, p. 185–198.

—— 1922, Metalliferous lodes in southern Seward Peninsula: U.S. Geol. Survey Bull. 722, p. 163–261.

Chapin, Theodore, 1914a, Placer mining in the Yukon-Tanana region: U.S. Geol. Survey Bull. 592, p. 357–362.

—— 1914b, Placer mining on Seward Peninsula: U.S. Geol. Survey Bull. 592, p. 385–395.

—— 1914c, Lode developments on Seward Peninsula: U.S. Geol. Survey Bull 592, p. 397–407.

—— 1918, The Nelchina-Susitna region, Alaska: U.S. Geol. Survey Bull. 668, 67 p.

—— 1919, Tin deposits of the Ruby district: U.S. Geol. Survey Bull. 692, p. 337.

Chapman, R. M., Coats, R. R., and Payne, T. G., 1963, Placer tin deposits in central Alaska: U.S. Geol. Survey open-file rept., 53 p.

Chapman, R. M., and Foster, R. L., 1969, Lode mines and prospects in the Fairbanks district, Alaska: U.S. Geol. Survey Prof. Paper 625–D, p. D1–D25.

Clark, A, L., and Cobb, E. H., 1970, Metallic mineral resources map of the Talkeetna quadrangle, Alaska: U.S. Geol. Survey open-file rept., 5 p.

Clark, A. L., and Hawley, C. C., 1968, Reconnaissance geology, mineral occurrences, and geochemical anomalies of the Yentna district, Alaska: U.S. Geol. Survey open-file rept., 64 p.

Clark, S. H. B., and Foster, H. L., 1969a, Preliminary geologic map of the Eagle D–2 and D–3 quadrangles, Alaska: U.S. Geol. Survey open-file map.

—— 1969b, Analyses of stream-sediment, rock, and soil samples from a part of the Seventymile River area, Eagle quadrangle, Alaska: U.S. Geol. Survey open-file rept., 129 p.

Coats, R. R., 1944a, Occurrences of scheelite in the Solomon district, Seward Peninsula, Alaska: U.S. Geol. Survey open-file rept., 4 p.

—— 1944b, Graphite deposits on the north side of the Kigluaik Mountains, Seward Peninsula, Alaska: U.S. Geol. Survey open-file rept., 8 p.

—— 1944c, Lode scheelite deposits of the Nome area, Seward Peninsula, Alaska: U.S. Geol. Survey open-file rept., 6 p.

—— 1950, Volcanic activity in the Aleutian arc: U.S. Geol. Survey Bull. 974–B, p. 35–49.

—— 1956, Geology of northern Adak Island, Alaska: U.S. Geol. Survey Bull. 1028–C, p. 45–67.

Cobb, E. H., 1960a, Chromite, cobalt, nickel, and platinum occurrences in Alaska: U.S. Geol. Survey Mineral Inv. Resource Map MR–8, scale 1:2,500,000.

—— 1960b, Copper, lead, and zinc occurrences in Alaska: U.S. Geol. Survey Mineral Inv. Resource Map MR–9, scale 1:2,500,000.

—— 1960c, Molybdenum, tin, and tungsten occurrences in Alaska: U.S. Geol. Survey Mineral Inv. Resource Map MR–10, scale 1:2,500,000.

—— 1960d, Antimony, bismuth, and mercury occurrences in Alaska: U.S. Geol. Survey Mineral Inv. Resource Map MR–11, scale 1:2,500,000.

—— 1962, Lode gold and silver occurrences in Alaska: U.S. Geol. Survey Mineral Inv. Resource Map MR–32, scale 1:2,500,000.

—— 1964a, Placer gold occurrences in Alaska: U.S. Geol. Survey Mineral Inv. Resource Map MR–38, scale 1:2,500,000.

—— 1964b, Iron occurrences in Alaska: U.S. Geol. Survey Mineral Inv. Resource Map MR–40, scale 1:2,500,000.

—— 1964c, Industrial minerals and construction materials occurrences in Alaska: U.S. Geol. Survey Mineral Inv. Resource Map MR–41, scale 1:2,500,000.

—— 1967a, compiler, Metallic mineral resources map of the Bendeleben quadrangle, Alaska: U.S. Geol. Survey open-file rept., 9 p.

—— 1967b, compiler, Metallic mineral resources map of the Big Delta quadrangle, Alaska: U.S. Geol. Survey open-file rept., 4 p.

—— 1967c, compiler, Metallic mineral resources map of the Candle quadrangle, Alaska: U.S. Geol. Survey open-file rept., 5 p.

—— 1967d, compiler, Metallic mineral resources map of the Chandalar quadrangle, Alaska: U.S. Geol. Survey open-file rept., 5 p.

—— 1967e, compiler, Metallic mineral resources map of the Charley River quadrangle, Alaska: U.S. Geol. Survey open-file rept., 3 p.

—— 1967f, compiler, Metallic mineral resources map of the Circle quadrangle, Alaska: U.S. Geol. Survey open-file rept., 6 p.

—— 1967g, compiler, Metallic mineral resources map of the Eagle quadrangle, Alaska: U.S. Geol. Survey open-file rept., 8 p.

—— 1967h, compiler, Metallic mineral resources map of the Fairbanks quadrangle, Alaska: U.S. Geol. Survey open-file rept., 8 p.

—— 1967i, compiler, Metallic mineral resources map of the Healy quadrangle, Alaska: U.S. Geol. Survey open-file rept., 5 p.

—— 1967j, compiler, Metallic mineral resources map of the Livengood quadrangle, Alaska: U.S. Geol. Survey open-file rept., 11 p.

—— 1967k, compiler, Metallic mineral resources map of the Mount Hayes quadrangle, Alaska: U.S. Geol. Survey open-file rept., 8 p.

—— 1967l, compiler, Metallic mineral resources map of the Solomon quadrangle, Alaska: U.S. Geol. Survey open-file rept., 10 p.

—— 1967m, compiler, Metallic mineral resources map of the Tanana quadrangle, Alaska: U.S. Geol. Survey open-file rept., 8 p.

—— 1967n, compiler, Metallic mineral resources map of the Wiseman quadrangle, Alaska: U.S. Geol. Survey open-file rept., 6 p.

—— 1968a, compiler, Metallic mineral resources map of the Ambler River quadrangle, Alaska: U.S. Geol. Survey open-file rept., 3 p.

—— 1968b, compiler, Metallic mineral resources map of the Baird Mountains quadrangle, Alaska: U.S. Geol. Survey open-file map, 2 p.

—————— 1968c, compiler, Metallic mineral resources map of the Bradfield Canal quadrangle, Alaska: U.S. Geol. Survey open-file rept., 3 p.

—————— 1968d, compiler, Metallic mineral resources map of the Craig quadrangle, Alaska: U.S. Geol. Survey open-file rept., 8 p.

—————— 1968e, compiler, Metallic mineral resources map of the Dixon Entrance quadrangle, Alaska: U.S. Geol. Survey open-file rept., 2 p.

—————— 1968f, compiler, Metallic mineral resources maps of nine Alaska quadrangles (Holy Cross, Kotzebue, Melozitna, Norton Bay, Nulato, Prince Rupert, Survey Pass, Taku River, Unalakleet): U.S. Geol. Survey open-file rept., 16 p.

—————— 1968g, compiler, Metallic mineral resources map of the Hughes quadrangle, Alaska: U.S. Geol. Survey open-file rept., 3 p.

—————— 1968h, compiler, Metallic mineral resources map of the Iditarod quadrangle, Alaska: U.S. Geol. Survey open-file rept., 5 p.

—————— 1968i, compiler, Metallic mineral resources map of the Juneau quadrangle, Alaska: U.S. Geol. Survey open-file rept., 7 p.

—————— 1968j, compiler, Metallic mineral resources map of the Ketchikan quadrangle, Alaska: U.S. Geol. Survey open-file rept., 4 p.

—————— 1968k, compiler, Metallic mineral resources map of the McGrath quadrangle, Alaska: U.S. Geol. Survey open-file map, 3 p.

—————— 1968l, compiler, Metallic mineral resources map of the Medfra quadrangle, Alaska: U.S. Geol. Survey open-file rept. 3 p.

—————— 1968m, compiler, Metallic mineral resources map of the Mount Fairweather quadrangle, Alaska: U.S. Geol. Survey open-file rept. 6 p.

—————— 1968n, compiler Metallic mineral resources map of the Nome quadrangle, Alaska: U.S. Geol. Survey open-file rept., 12 p.

—————— 1968o, compiler, Metallic mineral resources map of the Ophir quadrangle, Alaska: U.S. Geol. Survey open-file rept., 3 p.

—————— 1968p, compiler, Metallic mineral resources map of the Petersburg quadrangle, Alaska: U.S. Geol. Survey open-file rept., 3 p.

—————— 1968q, compiler, Metallic mineral resources map of the Port Alexander quadrangle, Alaska: U.S. Geol. Survey open-file rept., 3 p.

—————— 1968r, compiler, Metallic mineral resources map of the Ruby quadrangle, Alaska: U.S. Geol. Survey open-file rept., 5 p.

—————— 1968s, compiler, Metallic mineral resources map of the Shungnak quadrangle, Alaska: U.S. Geol. Survey open-file rept., 3 p.

—————— 1968t, compiler, Metallic mineral resources map of the Sitka quadrangle, Alaska: U.S. Geol. Survey open-file rept., 5 p.

—————— 1968u, compiler, Metallic mineral resources map of the Skagway quadrangle, Alaska: U.S. Geol. Survey open-file rept., 3 p.

—————— 1968v, compiler, Metallic mineral resources map of the Sumdum quadrangle, Alaska: U.S. Geol. Survey open-file rept., 2 p.

—————— 1969a, compiler, Metallic mineral resources maps of eleven Alaska quadrangles [Afognak, Beaver, Bering Glacier, Bettles, Karluk, Kenai, Middleton Island, Taylor Mountains, Trinity Islands, Tyonek, Yakutat]: U.S. Geol. Survey open-file rept., 20 p.

—————— 1969b, compiler, Metallic mineral resources map of the Cordova quadrangle, Alaska: U.S. Geol. Survey open-file rept., 4 p.

—————— 1969c, compiler, Metallic mineral resources map of the Hagemeister Island quadrangle, Alaska: U.S. Geol. Survey open-file rept. 3 p.

—————— 1969d, compiler, Metallic mineral resources map of the Kodiak quadrangle, Alaska: U.S. Geol. Survey open-file rept. 3 p.

———— 1969e, compiler, Metallic mineral resources map of the Seldovia quadrangle, Alaska: U.S. Geol. Survey open-file map, 4 p.

———— 1969f, compiler, Metallic mineral resources map of the Sleetmute quadrangle, Alaska: U.S. Geol. Survey open-file rept., 4 p.

———— 1969g, compiler, Metallic mineral resources map of the Talkeetna Mountains quadrangle, Alaska: U.S. Geol. Survey open-file rept., 4 p.

———— 1970a, compiler, Metallic mineral resources map of the Bethel quadrangle, Alaska: U.S. Geol. Survey open-file rept., 3 p.

———— 1970b, compiler, Metallic mineral resources maps of seven Alaska quadrangles: Chignik, Cold Bay, Dillingham, Lake Clark, Naknek, Port Moller, Unalaska: U.S. Geol. Survey open-file rept., 14 p.

Cobb, E. H., and Condon, W. H., 1970, compilers, Metallic mineral resources map of the Goodnews quadrangle, Alaska: U.S. Geol. Survey open-file map, 3 p.

Cobb, E. H., and Kachadoorian, Reuben, 1961, Index of metallic and nonmetallic mineral deposits of Alaska compiled from published reports of Federal and State agencies through 1959: U.S. Geol. Survey Bull. 1139, 363 p.

Cobb, E. H., and Matson, N. A., Jr., 1969, compilers, Metallic mineral resources map of the Anchorage quadrangle, Alaska: U.S. Geol. Survey open-file rept., 11 p.

Cobb, E. H., and Richter, D. H., 1967, compilers, Metallic mineral resources map of the Seward and Blying Sound quadrangles, Alaska: U.S. Geol. Survey open-file rept., 13 p.

Cobb, E. H., and Sainsbury, C. L., 1968, compilers, Metallic mineral resources map of the Teller quadrangle, Alaska: U.S. Geol. Survey open-file report, 8 p.

Cobb, E. H., Wanek, A. A., Grantz, Arthur, and Carter, Claire, 1968, Summary report on the geology and mineral resources of the Bering Sea, Bogoslof, Simeonof, Semidi, Tuxedni, St. Lazaria, Hazy Islands, and Forrester Island National Wildlife Refuges, Alaska: U.S. Geol. Survey Bull. 1260-K, p. K1-K28.

Collier, A. J., 1902, A reconnaissance of the northwestern portion of Seward Peninsula, Alaska: U.S. Geol. Survey Prof. Paper 2, 70 p.

———— 1905, Recent development of Alaskan tin deposits: U.S. Geol. Survey Bull. 259, p. 120-127.

Collier, A. J., Hess, F. L., Smith, P. S., and Brooks, A. H., 1908, The gold placers of parts of Seward Peninsula, Alaska, including the Nome, Council, Kougarok, Port Clarence, and Goodhope precincts: U.S. Geol. Survey Bull. 328, 343 p.

Coonrad, W. L., 1957, Geologic reconnaissance in the Yukon-Kuskokwim delta region, Alaska: U.S. Geol. Survey Misc. Geol. Inv. Map I-223, scale 1:500,000.

Coulter, H. W., Hopkins, D. M., Karlstrom, T. N. V., Péwé, T. L., Wahrhaftig, Clyde, and Williams, J. R., 1965, Map showing extent of glaciations in Alaska: U.S. Geol. Survey Misc. Geol. Inv. Map I-415, scale 1:2,500,000.

Davis, J. A., 1923, The Kantishna region, Alaska, in Stewart, B. D., Annual report of the mine inspector to the Governor of Alaska, 1922: Juneau, Alaska, p. 113-134.

Detterman, R. L., and Cobb, E. H., 1969, compilers, Metallic mineral resources map of the Iliamna quadrangle, Alaska: U.S. Geol. Survey open-file rept., 3 p.

Detterman, R. L., and Reed, B. L., 1968, Geology of the Iliamna quadrangle, Alaska: U.S. Geol. Survey open-file map.

Drewes, Harald, Fraser, G. D., Snyder, G. L., and Barnett, H. F., Jr., 1961, Geology of Unalaska Island and adjacent insular shelf, Alaska: U.S. Geol. Survey Bull. 1028-S, p. 583-676.

Dutro, J. T., Jr., and Payne, T. G., 1957, Geologic map of Alaska: U.S. Geol. Survey, scale 1:2,500,000.

Eakin, H. M., 1912, The Rampart and Hot Springs regions: U.S. Geol. Survey Bull. 520, p. 271-286.

——— 1913a, A geologic reconnaissance of a part of the Rampart quadrangle, Alaska: U.S. Geol. Survey Bull. 535, 38 p.

——— 1913b, Gold placers of the Innoko-Iditarod region: U.S. Geol. Survey Bull. 542, p. 293-303.

——— 1914a, The Iditarod-Ruby region, Alaska: U.S. Geol. Survey Bull. 578, 45 p.

——— 1914b, Placer mining in the Ruby district: U.S. Geol. Survey Bull. 592, p. 363-369.

——— 1914c, Mineral resources of the Yukon-Koyukuk region: U.S. Geol. Survey Bull. 592, p. 371-384.

——— 1915a, Tin mining in Alaska: U.S. Geol. Survey Bull. 622, p. 81-94.

——— 1915b, Mining in the Fairbanks district: U.S. Geol. Survey Bull. 622, p. 229-238.

——— 1915c, Placer mining in Seward Peninsula: U.S. Geol. Survey Bull. 622, p. 366—373.

——— 1916, The Yukon-Koyukuk region, Alaska: U.S. Geol. Survey Bull. 631, 88 p.

——— 1918, The Cosna-Nowitna region, Alaska: U.S. Geol. Survey Bull. 667, 54 p.

——— 1919, The Porcupine gold placer district, Alaska: U.S. Geol. Survey Bull. 699, 29 p.

Ellsworth, C. E., 1912, Placer mining in the Fairbanks and Circle districts: U.S. Geol. Survey Bull. 520, p. 240-245.

Ellsworth, C. E., and Davenport, R. W., 1913, Placer mining in the Yukon-Tanana region: U.S. Geol. Survey Bull. 542, p. 203-222.

Ellsworth, C. E., and Parker, G. L., 1911, Placer mining in the Yukon-Tanana region: U.S. Geol. Survey Bull. 480, p. 153-172.

Ferrians, O. J., Jr., 1965, Permafrost map of Alaska: U.S. Geol. Survey Misc. Geol. Inv. Map I-445, scale 1:2,500,000.

Ferrians, O. J., Jr., Kachadoorian, Reuben, and Greene, G. W., 1969, Permafrost and related engineering problems in Alaska: U.S. Geol. Survey Prof. Paper 678, 37 p.

Foster, H. L., 1967, Geology of the Mount Fairplay area, Alaska: U.S. Geol. Survey Bull. 1241-B, p. B1-B18.

——— 1968, Reconnaissance geologic map of the Tanacross quadrangle, Alaska: U.S. Geol. Survey open-file report, 13 p.

——— 1969a, Reconnaissance geology of the Eagle A-1, and A-2 quadrangles, Alaska: U.S. Geol. Survey Bull. 1271-G, p. G1-G30.

——— 1969b, Asbestos occurrence in the Eagle C-4 quadrangle, Alaska: U.S. Geol. Survey Circ. 611, 7 p.

Foster, H. L., and Clark, S. H. B., 1970, Geochemical and geologic reconnaissance of a part of the Fortymile area, Alaska: U.S. Geol. Survey
Bull. 1312-M, p. M1-M29.

Foster, H. L., and Keith, T. C., 1968, Preliminary geologic map of the Eagle
B-1 and C-1 quadrangles, Alaska: U.S. Geol. Survey open-file map.

—— 1969, Geology along the Taylor Highway, Alaska: U.S. Geol. Survey
Bull. 1281, 36 p.

Foster, R. L., 1966, The petrology and structure of the Amy Dome area,
Tolovana mining district, east-central Alaska: Columbia, Missouri Univ.
Ph.D. thesis, 227 p.

—— 1969, Nickeliferous serpentinites near Beaver Creek, east-central
Alaska, in U.S. Geological Survey, Some shorter mineral resource investigations in Alaska: U.S. Geol. Survey Circ. 615, p. 2-4.

Foster, R. L., and Chapman, R. M., 1967, Location and description of lode
prospects in the Livengood area, east-central Alaska: U.S. Geol. Survey
open-file rept., 3 p.

Freeman, V. L., 1963, Examination of uranium prospects, 1956, in U.S. Geological Survey, Contributions to economic geology of Alaska: U.S. Geol.
Survey Bull. 1155, p. 29-33.

Fritts, C. E., 1969, Geology and geochemistry in the southeastern part of
the Cosmos Hills, Shungnak D-2 quadrangle, Alaska: Alaska Div. Mines
and Geology Geol. Rept. 37, 35 p.

Gates, G. O., and Gryc, George, 1963, Structure and tectonic history of
Alaska, in Childs, O. E., and Beebe, B. W., eds., The backbone of the
Americas, a symposium: Am. Assoc. Petroleum Geologists Mem. 2, p.
264-277.

Gault, H. R., 1953, Candle Creek area, 1945, in Gault, H. R., and others,
1953, Reconnaissance for radioactive deposits in the northeastern part
of the Seward Peninsula, Alaska, 1945-47 and 1951: U.S. Geol. Survey
Circ. 250, p. 11-14.

Gault, H. R., Black, R. F., and Lyons, J. B., 1953, Sweepstakes Creek area,
1945, in Gault, H. R., and others, 1953, Reconnaissance for radioactive
deposits in the northeastern part of the Seward Peninsula, Alaska,
1945-47 and 1951: U.S. Geol. Survey Circ. 250, p. 1-10.

Gibson, T. M., 1911, Pay streaks at Nome: Mining Sci. Press, v. 102, p. 424-
427, 462-467.

Grant, U.S., 1909, Gold on Prince William Sound: U.S. Geol. Survey Bull.
379, p. 97.

Grant, U. S., and Higgins, D. F., 1910, Reconnaissance of the geology and
mineral resources of Prince William Sound, Alaska: U.S. Geol. Survey
Bull. 443, 89 p.

Grantz, Arthur, 1956, Possible origin of the placer gold deposits of the
Nelchina area, Alaska [abs.]: Geol. Soc. America Bull., v. 67, no. 12,
p. 1807.

—— 1966, Strike-slip faults in Alaska: U.S. Geol. Survey open-file rept.,
82 p.

Grantz, Arthur, Thomas, Herman, Stern, T. W., and Sheffey, N. B., 1963,
Potassium-argon and lead-alpha ages for stratigraphically bracketed
plutonic rocks in the Talkeetna Mountains, Alaska, in Short papers in
geology and hydrology: U.S. Geol. Survey Prof. Paper 475-B, p. B56-
B59.

Hanson, B. M., 1957, Middle Permian limestone on Pacific side of Alaska Peninsula: Am. Assoc. Petroleum Geologists Bull., v. 41, p. 2376–2378.

Harrington, G. L., 1918, The Anvik-Andreafski region, Alaska (including the Marshall district): U.S. Geol. Survey Bull. 683, 70 p.

——— 1919a, The gold and platinum placers of the Tolstoi district: U.S. Geol. Survey Bull. 692, p. 339–351.

——— 1919b, The gold and platinum placers of the Kiwalik-Koyuk region: U.S. Geol. Survey Bull. 692, p. 369–400.

——— 1921, Mineral resources of the Goodnews Bay region: U.S. Geol. Survey Bull. 714, p. 207–228.

Heide, H. E., and Rutledge, F. A., 1949, Investigation of Potato Mountain tin placer deposits, Seward Peninsula, northwestern Alaska: U.S. Bur. Mines Rept. Inv. 4418, 21 p.

Heide, H. E., and Sanford, R. S., 1948, Churn drilling at Cape Mountain tin placer deposits, Seward Peninsula, Alaska: U.S. Bur. Mines Rept. Inv. 4345, 14 p.

Henshaw, F. F., 1909, Mining in the Fairhaven precinct: U.S. Geol. Survey Bull. 379, p. 355–369.

——— 1910, Mining in Seward Peninsula: U.S. Geol. Survey Bull. 442, p. 353–371.

Herreid, Gordon, 1965a, Geology of the Bluff area, Solomon quadrangle, Seward Peninsula, Alaska: Alaska Div. Mines and Minerals Geol. Rept. 10, 21 p.

——— 1965b, Geology of the Omilak-Otter Creek area, Bendeleben quadrangle, Seward Peninsula, Alaska: Alaska Div. Mines and Minerals Geol. Rept. 11, 12 p.

——— 1965c, Geology of the Bear Creek area, Seward Peninsula, Candle quadrangle, Alaska: Alaska Div. Mines and Minerals Geol. Rept. 12, 16 p.

——— 1968, Progress report on the geology and geochemistry of the Sinuk area, Seward Peninsula, Alaska: Alaska Div. Mines and Minerals Geol. Rept. 29, 13 p.

——— 1969, Geology and geochemistry, Sithylemenkat Lake area, Bettles quadrangle, Alaska: Alaska Div. Mines and Geology Geol. Rept. 35, 22 p.

Hess, F. L., 1906, The York tin region: U.S. Geol. Survey Bull. 284, p. 145–157.

——— 1908, Placers of the Rampart region, in Prindle, L. M., The Fairbanks and Rampart quadrangles, Yukon-Tanana region, Alaska, with a section on Placers of the Rampart region, by F. L., Hess, and a paper on Water supply of the Fairbanks region, by C. C. Covert: U.S. Geol. Survey Bull. 337, p. 64–98.

Hill, J. M., 1933, Lode deposits of the Fairbanks district, Alaska: U.S. Geol. Survey Bull. 849–B, p. 29–163.

Hoare, J. M., 1961, Geology and tectonic setting of lower Kuskokwim-Bristol Bay region, Alaska: Am. Assoc. Petroleum Geologists Bull., v. 45, no. 5, p. 594–611.

Hoare, J. M., and Cobb, E. H., 1970, Metallic mineral resources map of the Russian Mission quadrangle, Alaska: U.S. Geol. Survey open-file rept. 5 p.

Hoare, J. M., and Condon, W. H., 1966, Geologic map of the Kwiguk and Black quadrangles, western Alaska: U.S. Geol. Survey Misc. Geol. Inv. Map I-469, 7 p, scale 1:250,000.

—— 1968, Geologic map of the Hooper Bay quadrangle, Alaska: U.S. Geol. Survey Misc. Geol. Inv. Map I–523, scale 1:250,000.

—— 1971a, Geologic map of the Marshall quadrangle, Alaska: U.S. Geol. Survey Misc. Geol. Inv. Map I–668, scale 1:250,000.

—— 1971b, Geologic map of the St. Michael quadrangle, Alaska: U.S. Geol. Survey Misc. Geol. Inv. Map I–682, scale 1:250,000.

Hoare, J. M., Condon, W. H., Cox, Allan, and Dalrymple, G. B., 1968, Geology, paleomagnetism, and potassium-argon ages of basalts from Nunivak Island, Alaska: Geol. Soc. America Mem. 116, p. 377–413.

Hoare, J. M., and Coonrad, W. L., 1959a, Geology of the Bethel quadrangle, Alaska: U.S. Geol. Survey Misc. Geol. Inv. Map I–285, scale 1:250,000.

—— 1959b, Geology of the Russian Mission quadrangle, Alaska: U.S. Geol. Survey Misc. Geol. Inv. Map I–292, scale 1:250,000.

—— 1961, Geologic map of the Goodnews quadrangle, Alaska: U.S. Geol. Survey Misc. Geol. Inv. Map I–339, scale 1:250,000.

Hopkins, D. M., 1963, Geology of the Imuruk Lake area, Seward Peninsula, Alaska: U.S. Geol. Survey Bull. 1141–C, p. C1–C101.

Hopkins, D. M., ed., 1967, The Bering Sea land bridge: Stanford, Calif., Stanford Univ. Press, 495 p.

Hopkins, D. M., MacNeil, F. S., and Leopold, E. B., 1960, The coastal plain at Nome—a late Cenozoic type section for the Bering Strait region, in Chronology and climatology of the Quaternary: Internat. Geol. Cong., 21st, Copenhagen 1960, Proc., pt. 4, p. 46–57.

Hummel, C. L., 1960, Structural geology and structural control of mineral deposits near Nome, Alaska, in Short papers in the geological sciences: U.S. Geol. Survey Prof. Paper 400–B, p. B33–B35.

—— 1961, Regionally metamorphosed metalliferous contact-metasomatic deposits near Nome, Alaska, in Short papers in the geologic and hydrologic sciences: U.S. Geol. Survey Prof. Paper 424–D, p. D198–D199.

—— 1962a, Preliminary geologic map of the Nome C–1 quadrangle, Seward Peninsula, Alaska: U.S. Geol. Survey Mineral Inv. Field Studies Map MF–247, scale 1:63,360.

—— 1962b, Preliminary geologic map of the Nome D–1 quadrangle, Seward Peninsula, Alaska: U.S. Geol. Survey Mineral Inv. Field Studies Map MF–248, scale 1:63,360.

Jasper, M. W., 1961, Bonanza Creek placers: Alaska Div. Mines and Minerals Rept. 1961, p. 58–64.

—— 1962, Willow Creek gold district activity, Anchorage quadrangle: Alaska Div. Mines and Minerals Rept. 1962, p. 75–84.

—— 1967a, Geochemical investigations, Willow Creek southerly to Kenai Lake region, south central Alaska: Alaska Div. Mines and Minerals Geochem. Rept. 14, 47 p.

—— 1967b, Geochemical investigations along the Valdez to Chitina highway in southcentral Alaska, 1966: Alaska Div. Mines and Minerals Geochem. Rept. 15, 19 p.

Joesting, H. R., 1942, Strategic mineral occurrences in interior Alaska: Alaska Dept. Mines Pamph. 1, 46 p.

—— 1943, Supplement to Pamphlet No. 1—Strategic mineral occurrences in interior Alaska: Alaska Dept. Mines Pamph. 2, 28 p.

Johnson, B. L., 1910, Occurrence of wolframite and cassiterite in the gold placers of Deadwood Creek, Birch Creek district: U.S. Geol. Survey Bull. 442, p. 246–250.

—— 1915, The gold and copper deposits of the Port Valdez district: U.S. Geol. Survey Bull. 622, p. 140–188.

Jones, D. L., and MacKevett, E. M., Jr., 1969, Summary of Cretaceous stratigraphy in part of the McCarthy quadrangle, Alaska: U.S. Geol. Survey Bull. 1274–K, p. K1–K19.

Kelly, T. E., 1963, Geology and hydrocarbons in Cook Inlet basin, Alaska, *in* Childs, O. E., and Beebe, B. W., eds., The backbone of the Americas, a symposium: Am. Assoc. Petroleum Geologists Mem. 2, p. 278–296.

Killeen, P. L., and Ordway, R. J., 1955, Radioactivity investigations at Ear Mountain, Seward Peninsula, Alaska, 1945: U.S. Geol. Survey Bull. 1024–C, p. 59–94.

Killeen, P. L., and White, M. G., 1953, South Fork of Quartz Creek, 1946, *in* Gault, H. R., and others, 1953, Reconnanssace for radioactive deposits in the northeastern part of the Seward Peninsula, Alaska, 1945–47 and 1951: U.S. Geol. Survey Circ. 250, p. 15–20.

Kimball, A. L., 1969a, Reconnaissance of Tatonduk River red beds: U.S. Bur. Mines open-file rept., 11 p.

—— 1969b, Reconnaissance sampling of decomposed monzonite for gold near Flat, Alaska: U.S. Bur. Mines open-file rept., 39 p.

Knopf, Adolph, 1908, Geology of the Seward Peninsula tin deposits, Alaska: U.S. Geol. Survey Bull. 358, 71 p.

Lathram, E. H., 1965, Preliminary geologic map of northern Alaska: U.S. Geol. Survey open-file map.

MacKevett, E. M., Jr., Brew, D. A., Hawley, C. C., Huff, L. C., and Smith, J. G., 1967, Mineral resources of Glacier Bay National Monument, Alaska: U.S. Geol. Survey open-file rept., 176 p.

MacKevett, E. M., Jr., and Cobb, E. H., 1969, compilers, Metallic mineral resources map of the McCarthy quadrangle, Alaska: U.S. Geol. Survey open-file rept., 8 p.

MacKevett, E. M., Jr., and Smith, J. G., 1968, Distribution of gold, copper, and some other metals in the McCarthy B–4 and B–5 quadrangles, Alaska: U.S. Geol. Survey Circ. 604, 25 p.

MacNeil, F. S., Wolfe, J. A., Miller, D. J., and Hopkins, D. M., 1961, Correlation of Tertiary formations of Alaska: Am Assoc. Petroleum Geollogists Bull., v. 45, no. 11, p. 1801–1809.

Maddren, A. G., 1910, The Innoko gold-placer district, Alaska, with accounts of the central Kuskokwim Valley and the Ruby Creek and Gold Hill placers: U.S. Geol. Survey Bull. 410, 87 p.

—— 1911, Gold placer mining developments in the Innoko-Iditarod region: U.S. Geol. Survey Bull. 480, p. 236–270.

—— 1913, The Koyukuk-Chandalar region, Alaska: U.S. Geol. Survey Bull. 532, 119 p.

—— 1914, Mineral deposits of the Yakataga district: U.S. Geol. Survey Bull. 592, p. 119–153.

—— 1915, Gold placers of the lower Kuskokwim, with a note on copper in the Russian Mountains: U.S. Geol. Survey Bull. 622, p. 292–360.

—— 1918, Gold placers near the Nenana coal field: U.S. Geol. Survey Bull. 662, p. 363–402.

—— 1919, The beach placers of the west coast of Kodiak Island, Alaska: U.S. Geol. Survey Bull. 692, p. 299–319.

Malone, Kevin, 1962, Mercury occurrences in Alaska: U.S. Bur. Mines Inf. Circ. 8131, 57 p.

Maloney, R. P., 1962, Investigation of mercury-antimony deposits near Flat, Yukon River region, Alaska: U.S. Bur. Mines Rept. Inv. 5991, 44 p.

Mangus, M. D., 1953, Regional interpretation of the geology of the Kongakut-Firth Rivers area, Alaska: U.S. Geol. Survey Inv. Naval Petroleum Reserve No. 4 and adjacent areas Spec. Rept. 43, 24 p.

Martin, G. C., 1913, Mineral deposits of Kodiak and the neighboring islands: U.S. Geol. Survey Bull. 542, p. 125–136.

———— 1919, The Alaskan mining industry in 1917: U.S. Geol. Survey Bull. 692, p. 11–42.

———— 1920, The Alaskan mining industry in 1918: U.S. Geol. Survey Bull. 712, p. 11–52.

Martin, G. C., Johnson, B. L., and Grant, U.S., 1915, Geology and mineral resources of Kenai Peninsula, Alaska: U.S. Geol. Survey Bull. 587, 243 p.

Martin, G. C., and Mertie, J. B., Jr., 1914, Mineral resources of the upper Matanuska and Nelchina valleys: U.S. Geol. Survey Bull. 592, p. 273–299.

Matson, N. A., Jr., 1969a, compiler, Metallic mineral resources map of the Gulkana quadrangle, Alaska: U.S. Geol. Survey open-file rept., 4 p.

———— 1969b, compiler, Metallic mineral resources map of the Nabesna quadrangle, Alaska: U.S. Geol. Survey open-file rept., 6 p.

———— 1969c, compiler, Metallic mineral resources map of the Valdez quadrangle, Alaska: U.S. Geol. Survey open-file rept., 11 p.

Mendenhall, W. C., 1900, A reconnaissance from Resurrection Bay to the Tanana River, Alaska, in 1898: U.S. Geol. Survey 20th Ann. Rept., pt. 7, p. 265–340.

———— 1901, A reconnaissance in the Norton Bay region, Alaska, in 1900: U.S. Geol. Survey Spec. Pub., p. 181–222.

———— 1902, Reconnaissance from Fort Hamlin to Kotzebue Sound, Alaska, by way of Dall, Kanuti, Allen, and Kowak Rivers: U.S. Geol. Survey Prof. Paper 10, 68 p.

———— 1905, Geology of the central Copper River region, Alaska: U.S. Geol. Survey Prof. Paper 41, 133 p.

Mendenhall, W. C., and Schrader, F. C., 1903, The mineral resources of the Mount Wrangell district, Alaska: U.S. Geol. Survey Prof. Paper 15, 71 p.

Mertie, J. B., Jr., 1918a, The gold placers of the Tolovana district: U.S. Geol. Survey Bull. 662, p. 221–277.

———— 1918b, Placer mining on Seward Peninsula: U.S. Geol. Survey Bull. 662, p. 451–458.

———— 1919, Platinum-bearing placers of the Kahiltna Valley: U.S. Geol. Survey Bull. 692, p. 233–264.

———— 1925, Geology and gold placers of the Chandalar district: U.S. Geol. Survey Bull. 773, p. 215–263.

———— 1930, The Chandalar-Sheenjek district: U.S. Geol. Survey Bull. 810, p. 87–139.

———— 1932, Mining in the Circle district: U.S. Geol. Survey Bull. 824, p. 155–172.

———— 1933, Notes on the geography and geology of Lituya Bay: U.S. Geol. Survey Bull. 836, p. 117–135.

———— 1934, Mineral deposits of the Rampart and Hot Springs districts, Alaska: U.S. Geol. Survey Bull. 844–D, p. 163–226.

———— 1936, Mineral deposits of the Ruby-Kuskokwim region, Alaska: U.S. Geol. Survey Bull. 864–C, p. 115–255.

———— 1937a, The Kaiyuh Hills, Alaska: U.S. Geol. Survey Bull. 868–D, 145–178.

———— 1937b, The Yukon-Tanana region, Alaska: U.S. Geol. Survey Bull. 872, 276 p.

———— 1938a, Gold placers of the Fortymile, Eagle, and Circle districts, Alaska: U.S. Geol. Survey Bull. 897–C, p. 133–261.

———— 1938b, The Nushagak district, Alaska: U.S. Geol. Survey Bull. 903, 96 p.

———— 1940, The Goodnews platinum deposits, Alaska: U.S. Geol. Survey Bull. 918, 97 p.

———— 1942, Tertiary deposits of the Eagle-Circle district, Alaska: U.S. Geol. Survey Bull. 917–D, p. 213–264.

———— 1969, Economic geology of the platinum minerals: U.S. Geol. Survey Prof. Paper 630, 120 p.

Mertie, J. B., Jr., and Harrington, G. L.. 1916, Mineral resources of the Ruby-Kuskokwim region: U.S. Geol. Survey Bull. 642, p. 223–266.

———— 1924, The Ruby-Kuskokwim region, Alaska: U.S. Geol. Survey Bull. 754, 129 p.

Miller, D. J., 1946, Copper deposits of the Nizina district, Alaska: U.S. Geol. Survey Bull. 947–F, p. 93–120.

———— 1957, Geology of the southeastern part of the Robinson Mountains, Yakataga district, Alaska: U.S. Geol. Survey Oil and Gas Inv. Map OM–187, scale 1:63,360.

Miller, T. P., and Ferrians, O. J., Jr., 1968, Suggested areas for prospecting in the central Koyukuk River region, Alaska: U.S. Geol. Survey Circ. 570, 12 p.

Miller, T. P., Patton, W. W., Jr., and Lanphere, M. A., 1966, Preliminary report on a plutonic belt in west-central Alaska, in Geological Survey research 1966: U.S. Geol. Survey Prof. Paper 550–D, p. D158–D162.

Mining World, 1941, Mining through Arctic ice: Mining World, v. 3, no. 3, p. 3–7.

Moffit, F. H., 1905, The Fairhaven gold placers, Seward Peninsula, Alaska: U.S. Geol. Survey Bull. 247, 85 p.

———— 1906a, Gold fields of the Turnagain Arm region: U.S. Geol. Survey Bull. 277, p. 7–52.

———— 1906b, Gold mining on Seward Peninsula: U.S. Geol. Survey Bull. 284, p. 132–144.

———— 1912, Headwater regions of Gulkana and Susitna Rivers, Alaska, with accounts of the Valdez Creek and Chistochina placer districts: U.S. Geol. Survey Bull. 498, 82 p.

———— 1913, Geology of the Nome and Grand Central quadrangles, Alaska: U.S. Geol. Survey Bull. 533, 140 p.

———— 1914, Geology of the Hanagita-Bremner region, Alaska: U.S. Geol. Survey Bull. 576, 56 p.

———— 1915, The Broad Pass region, Alaska, with sections on Quaternary deposits, Igneous rocks. and Glaciation by Joseph E. Pogue: U.S. Geol. Survey Bull. 608, 80 p.

—— 1916, Mineral resources of the upper Chitina Valley: U.S. Geol. Survey Bull. 642, p. 129–136.

—— 1918, Mining in the lower Copper River basin: U.S. Geol. Survey Bull. 662, p. 155–182.

—— 1927, Mineral industry in Alaska in 1925: U.S. Geol. Survey Bull. 792, p. 1–39.

—— 1933, Mining development in the Tatlanika and Totatlanika basins: U.S. Geol. Survey Bull. 836, p. 339–345.

—— 1938a, Geology of the Chitina Valley and adjacent area, Alaska: U.S. Geol. Survey Bull. 894, 137 p.

—— 1938b, Geology of the Slana-Tok district, Alaska: U.S. Geol. Survey Bull. 904, 54 p.

—— 1941, Geology of the upper Tetling River district, Alaska: U.S. Geol. Survey Bull. 917–B, p. 115–157.

—— 1942, Geology of the Gerstle River district, Alaska, *with a report on* Black Rapids Glacier, Alaska, by Fred H. Moffit: U.S. Geol. Survey Bull. 926–B, p. 107–160.

—— 1943, Geology of the Nutzotin Mountains, Alaska, *with a section on* Igneous rocks, by Russell G. Wayland: U.S. Geol. Survey Bull. 933–B, p. 103–174.

—— 1944, Mining in the northern Copper River region, Alaska: U.S. Geol. Survey Bull. 943–B, p. 25–47.

—— 1954a, Geology of the eastern part of the Alaska Range and adjacent area: U.S. Geol. Survey Bull. 989–D, p. 63–218.

—— 1954b, Geology of the Prince William Sound region, Alaska: U.S. Geol. Survey Bull. 989–E, p. 225–310.

Moffit. F. H., and Capps, S. R., 1911, Geology and mineral resources of the Nizina district, Alaska: U.S. Geol. Survey Bull. 448, 111 p.

Moore, G. W., 1967, Preliminary geologic map of Kodiak Island and vicinity, Alaska: U.S. Geol. Survey open-file map.

Moxham, R. M., 1954, Reconnaissance for radioactive deposits in the Manley Hot Springs-Rampart district, east-central Alaska, 1948: U.S. Geol. Survey Circ. 317, 6 p.

Moxham, R. M., and Nelson, A. E., 1952, Reconnaissance for radioactive deposits in south-central Alaska, 1947–49: U.S. Geol. Survey Circ. 184, 14 p.

Moxham, R. M., and West, W. S., 1953, Radioactivity investigations in the Serpentine-Kougarok area, Seward Peninsula, Alaska, 1946: U.S. Geol. Survey Circ. 265, 11 p.

Mulligan, J. J., 1959a, Tin placer and lode investigations, Ear Mountain area, Seward Peninsula, Alaska: U.S. Bur. Mines Rept. Inv. 5493, 53 p.

—— 1959b, Sampling stream gravels for tin, near York, Seward Peninsula, Alaska: U.S. Bur. Mines Rept. Inv. 5520, 25 p.

—— 1965a, Tin-lode investigations, Potato Mountain area, Seward Peninsula, Alaska: U.S. Bur. Mines Rept. Inv. 6587, 85 p.

—— 1965b, Examination of Hannum prospect, Fairhaven district, Seward Peninsula, Alaska: U.S. Bur. Mines open-file rept., 16 p.

—— 1966, Tin-lode investigations, Cape Mountain area, Seward Peninsula, Alaska; *with a section on* Petrography by W. L. Gnagy: U.S. Bur. Mines Rept. Inv. 6737, 43 p.

Mulligan, J. J., and Thorne, R. L., 1959, Tin-placer sampling methods and results, Cape Mountain district, Seward Peninsula, Alaska: U.S. Bur. Mines Inf. Circ. 7878, 69 p.

Nelson, A. E., West, W. S., and Matzko, J. J., 1954, Reconnaissance for radioactive deposits in eastern Alaska, 1952: U.S. Geol. Survey Circ. 348, 21 p.

Nelson, C. H., Hopkins, D. M., and Ness, Gordon, 1969, Interpreting complex relict and modern sediment patterns on the Bering Shelf: Geol. Soc. America Abstracts with Programs, 1969, pt. 7, p. 159.

Nelson, Hans, and Hopkins, D. M., 1969, Sedimentary processes and distribution of particulate gold in northern Bering Sea: U.S. Geol. Survey open-file rept., 59 p.

Overbeck, R. M., 1918, Lode deposits near the Nenana coal field: U.S. Geol. Survey Bull. 662, p. 351–362.

—— 1920, Placer mining in the Tolovana district: U.S. Geol. Survey Bull. 712, p. 177–184.

Paige, Sidney, and Knopf, Adolph, 1907a, Reconnaissance in the Matanuska and Talkeetna basins, with notes on the placers of the adjacent region: U.S. Geol. Survey Bull. 314, p. 104–125.

—— 1907b, Geologic reconnaissance in the Matanuska and Talkeetna basins, Alaska: U.S. Geol. Survey Bull. 327, 71 p.

Patton, W. W., Jr., 1966, Regional geology of the Kateel River quadrangle, Alaska: U.S. Geol. Survey Misc. Geol. Inv. Map I–437, scale 1:250,000.

—— 1967, Regional geologic map of the Candle quadrangle, Alaska: U.S. Geol. Survey Misc. Geol. Inv. Map I–492, scale 1:250,000.

Patton, W. W., Jr., and Csejtey, Béla, Jr., 1970, Analyses of stream-sediment samples from western St. Lawrence Island, Alaska: U.S. Geol. Survey open-file rept., 50 p.

—— 1971, Preliminary geologic investigations of western St. Lawrence Island, Alaska: U.S. Geol. Survey Prof. Paper 684–C, 15 p.

Patton, W. W., Jr., and Dutro, J. T., Jr., 1969, Preliminary report on the Paleozoic and Mesozoic sedimentary sequence on St. Lawrence Island, Alaska, in Geological Survey research 1969: U.S. Geol. Survey Prof. Paper 650–D, p. D138–D143.

Patton, W. W., Jr., and Hoare, J. M., 1968, The Kaltag fault, west-central Alaska, in Geological Survey research 1968: U.S. Geol. Survey Prof. Paper 600–D, p. D147–D153.

Patton, W. W., Jr., and Miller, T. P., 1966, Regional geologic map of the Hughes quadrangle, Alaska: U.S. Geol. Survey Misc. Geol. Inv. Map I–459, scale 1:250,000

—— 1968, Regional geologic map of the Selawik and southern Baird Mountains quadrangles, Alaska: U.S. Geol. Survey Misc. Geol. Inv. Map I–530, scale 1:250,000.

—— 1970, Preliminary geologic investigations in the Kanuti River region, Alaska: U.S. Geol. Survey Bull. 1312–J, p. J1–J10.

Patton, W. W., Jr., Miller, T. P., and Tailleur, I. L., 1968, Regional geologic map of the Shungnak and southern Ambler River quadrangles, Alaska: U.S. Geol. Survey Misc. Geol. Inv. Map I–554, scale 1:250,000.

Péwé, T. L., 1955, Origin of the upland silt near Fairbanks, Alaska: Geol. Soc. America Bull., v. 66, p. 699–724.

Péwé, T. L., Wahrhaftig, Clyde, and Weber, Florence, 1966, Geologic map of the Fairbanks quadrangle, Alaska: U.S. Geol. Survey Misc. Geol. Inv. Map I–455, scale 1:250,000.

Plafker, George, 1967, Geologic map of the Gulf of Alaska Tertiary province, Alaska: U.S. Geol. Survey Misc. Geol. Inv. Map I–484, scale 1:500,000.

Plafker, George, and MacNeil, F. S., 1966, Stratigraphic significance of Tertiary fossils from the Orca Group in the Prince William Sound region, Alaska, in Geological Survey research 1966: U.S. Geol. Survey Prof. Paper 550–B, p. B62–B68.

Porter, E. A., 1912, Placer mining in the Fortymile, Eagle, and Seventymile River districts: U.S. Geol. Survey Bull. 520, p. 211–218.

Prindle, L. M., 1905, The gold placers of the Fortymile, Birch Creek, and Fairbanks regions, Alaska: U.S. Geol. Survey Bull. 251, 89 p.

—— 1906, Yukon placer fields: U.S. Geol. Survey Bull. 284, p. 109–127.

—— 1907, The Bonnifield and Kantishna regions, Alaska: U.S. Geol. Survey Bull. 314, p. 205–226.

—— 1908a, The Fairbanks and Rampart quadrangles, Yukon-Tanana region, Alaska, with a section on Placers of the Rampart region, by F. L. Hess, and a paper on Water supply of the Fairbanks region, by C. C. Covert: U.S. Geol. Survey Bull. 337, 102 p.

—— 1908b, The Fortymile gold placer district: U.S. Geol. Survey Bull. 345, p. 187–197.

—— 1909, The Fortymile quadrangle, Yukon-Tanana region, Alaska: U.S. Geol. Survey Bull. 375, 52 p.

—— 1910a, Sketch of the geology of the northeast part of the Fairbanks quadrangle: U.S. Geol. Survey Bull. 442, p. 203–209.

—— 1910b, Auriferous quartz veins in the Fairbanks district: U.S. Geol. Survey Bull. 442, p. 210–229.

—— 1913a, A geologic reconnaissance of the Fairbanks quadrangle, Alaska, with a detailed description of Geology of the Fairbanks district, by L. M. Prindle and F. J. Katz, and an account of Lode mining near Fairbanks, by P. S. Smith: U.S. Geol. Survey Bull. 525, 220 p.

—— 1913b, A geologic reconnaissance of the Circle quadrangle, Alaska: U.S. Geol. Survey Bull. 538, 82 p.

Prindle, L. M., and Hess, F. L., 1906, The Rampart gold placer region, Alaska: U.S. Geol. Survey Bull. 280, 54 p.

Prindle, L. M., and Katz, F. J., 1909, The Fairbanks gold-placer region: U.S. Geol. Survey Bull. 379, p. 181–200.

—— 1913, Geology of the Fairbanks district, in Prindle, L. M., A geologic reconnaissance of the Fairbanks quadrangle, Alaska: U.S. Geol. Survey Bull. 525, p. 59–152.

Ransome, A. L., and Kerns, W. H., 1954, Names and definitions of regions, districts, and subdistricts in Alaska (Used by the Bureau of Mines in statistical and economic studies covering the mineral industry of the Territory): U.S. Bur. Mines Inf. Circ. 7679, 91 p.

Reed, B. L., 1968, Geology of the Lake Peters area, northeastern Alaska: U.S. Geol. Survey Bull. 1236, 132 p.

Reed, B. L., and Eberlein, G. D., 1972, Massive sulfide deposits near Shellabarger Pass, southern Alaska Range, Alaska: U.S. Geol. Survey Bull. 1342, 45 p.

Reed, B. L., and Elliott, R. L., 1968a, Lead, zinc, and silver deposits at Bowser Creek, McGrath A–2 quadrangle, Alaska: U.S. Geol. Survey Circ. 559, 17 p.

—— 1968b, Geochemical anomalies and metalliferous deposits between Windy Fork and Post River, southern Alaska Range: U.S. Geol. Survey

Circ. 569, 22 p.

—— 1970, Reconnaissance geologic map, analyses of bedrock and stream sediment samples, and an aeromagnetic map of parts of the southern Alaska Range: U.S. Geol. Survey open-file rept., 145 p.

Reed, I. McK., 1938, Upper Koyukuk region, Alaska: Alaska Dept. Mines unpub. rept., 169 p.

Reed, Irving, 1931a, Report on the placer deposits of the upper Kobuk goldfields: Alaska Dept. Mines unpub. rept., 33 p.

—— 1931b, Report on the placer deposits of the Squirrel River gold field: Alaska Dept. Mines unpub. rept., 15 p.

Reed, J. C., and Coats, R. R., 1942, Geology and ore deposits of the Chichagof mining district, Alaska: U.S. Geol. Survey Bull. 929, 148 p.

Reed, J. C., Jr., 1961, Geology of the Mount McKinley quadrangle, Alaska: U.S. Geol. Survey Bull. 1108–A, p. A1–A36.

Reiser, H. N., Lanphere, M. A., and Brosgé, W. P., 1965, Jurassic age of a mafic igneous complex, Christian quadrangle, Alaska, in Geological Survey research 1965: U.S. Geol. Survey Prof. Paper 525–C, p. C68–C71.

Reiser, H. N., and Tailleur, I. L., 1969, compilers, Preliminary geologic map of Mt. Michelson quadrangle, Alaska: U.S. Geol. Survey open-file map.

Richter, D. H., 1966, Geology of the Slana district, southcentral Alaska: Alaska Div. Mines and Minerals Geol. Rept. 21, 51 p.

—— 1967a, Geological and geochemical investigations in the Metal Creek area, Chugach Mountains, Alaska: Alaska Div. Mines and Minerals Geol. Rept. 25, 17 p.

—— 1967b, Geology of the upper Slana-Mentasta Pass area, southcentral Alaska: Alaska Div. Mines and Minerals Geol. Rept. 30, 25 p.

—— 1970, Geology and lode-gold deposits of the Nuka Bay area, Kenai Peninsula, Alaska: U.S. Geol. Survey Prof. Paper 625–B, p. B1–B16.

Richter, D. H., and Matson, N. A., Jr., 1968, Distribution of gold and some base metals in the Slana area, eastern Alaska Range, Alaska: U.S. Geol. Survey Circ. 593, 20 p.

Robertson, E. C., 1956, Magnetite deposits near Klukwan and Haines, Alaska: U.S. Geol. Survey open-file rept., 37 p.

Robinson, G. D., Wedow, Helmuth, Jr., and Lyons, J. B., 1955, Radioactivity investigations in the Cache Creek area, Yenta district, Alaska, 1945: U.S. Geol. Survey Bull. 1024–A, p. 1–23.

Rose, A. W., 1965a, Geology and mineral deposits of the Rainy Creek area, Mt. Hayes quadrangle, Alaska: Alaska Div. Mines and Minerals Geol. Rept. 14, 51 p.

—— 1965b, Geology and mineralization of the Midas mine and Sulphide Gulch areas near Valdez, Alaska: Alaska Div. Mines and Minerals Geol. Rept. 15, 21 p.

—— 1967, Geology of the upper Chistochina River area, Mt. Hayes quadrangle, Alaska: Alaska Div. Mines and Minerals Geol. Rept. 28, 41 p.

Rose, A. W., and Saunders, R. H., 1965, Geology and geochemical investigations near Paxson, northern Copper River Basin, Alaska: Alaska Div. Mines and Minerals Geol. Rept. 13, 35 p.

Ross, C. P., 1933, The Valdez Creek mining district, Alaska: U.S. Geol. Survey Bull. 849–H, p. 425–468.

Rossman, D. L., 1957, Ilmenite-bearing beach sands near Lituya Bay, Alaska: U.S. Geol. Survey open-file rept., 10 p.

———— 1963, Geology of the eastern part of the Mount Fairweather quadrangle, Glacier Bay, Alaska: U.S. Geol. Survey Bull. 1121-K, p. K1-K57.

Rutledge, F. A., 1948, Investigation of the Rainy Creek mercury prospect, Bethel district, Kuskokwim region, southwestern Alaska: U.S. Bur. Mines Rept. Inv. 4361, 7 p.

Sainsbury, C. L., 1957, Some pegmatite deposits in southeastern Alaska: U.S. Geol. Survey Bull. 1024-G, p. 141-161.

———— 1964, Geology of the Lost River mine area, Alaska: U.S. Geol. Survey Bull. 1129, 80 p.

———— 1967, Upper Pleistocene features in the Bering Strait area, in Geological Survey research 1967: U.S. Geol. Survey Prof. Paper 575-D, p. D203-D213.

———— 1969a, Geology and ore deposits of the central York Mountains, western Seward Peninsula, Alaska: U.S. Geol. Survey Bull. 1287, 101 p.

———— 1969b, The A. J. Collier thrust belt of the Seward Peninsula, Alaska: Geol. Soc. America Bull., v. 80, no. 12, p. 2595-2596.

———— 1969c, Geologic map of the Teller B-4 and southern part of the Teller C-4 quadrangles, western Seward Peninsula, Alaska: U.S. Geol. Survey Misc. Geol. Inv. Map I-572, scale 1:63,360.

Sainsbury, C. L., Kachadoorian, Reuben, Smith, T. E., and Todd, W. C., 1968, Cassiterite in gold placers at Humboldt Creek, Serpentine-Kougarok area, Seward Peninsula, Alaska: U.S. Geol. Survey Circ. 565, 7 p.

Sainsbury, C. L., and MacKevett, E. M., Jr., 1965, Quicksilver deposits of southwestern Alaska: U.S. Geol. Survey Bull. 1187, 89 p.

Sanders, R. H., 1965, A geochemical investigation in the Richardson area, Big Delta quadrangle, Alaska: Alaska Div. Mines and Minerals Geochem. Rept. 3, 12 p.

Scholl, D. W., Greene, H. G., Addicott, W. O., Evitt, W. R., Pierce, R. L., Mamay, S. H., and Marlow, M. S., 1969, Adak "Paleozoic" site, Aleutians —in fact of Eocene age [abs.]: Am. Assoc. Petroleum Geologists Bull., v. 53, p. 459.

Scholl, D. W., and Hopkins, D. M., 1969, Newly discovered Cenozoic basins, Bering Sea shelf, Alaska: Am. Asoc. Petroleum Geologists Bull., v. 53, p. 2067-2078.

Schrader, F. C., 1900, A reconnaissance of a part of Prince William Sound and the Copper River district, Alaska, in 1898: U.S. Geol. Survey 20th Ann. Rept., pt. 7, p. 341-423.

———— 1904, A reconnaissance in northern Alaska across the Rocky Mountains, along Koyukuk, John, Anaktuvuk, and Colville Rivers and the Arctic coast to Cape Lisburne, in 1901: U.S. Geol. Survey Prof. Paper 20, 139 p.

Schrader, F. C., and Brooks, A. H., 1900, Preliminary report on the Cape Nome gold region, Alaska: U.S. Geol. Survey Spec. Pub., 56 p.

Smith, J. G., and MacKevett, E. M., Jr., 1970, The Skolai Group in the McCarthy B-4, C-4, and C-5 quadrangles, Wrangell Mountains, Alaska: U.S. Geol. Survey Bull. 1274-Q, p. Q1-Q26.

Smith, P. S., 1907, Geology and mineral resources of Iron Creek: U.S. Geol. Survey Bull. 314, p. 157-163.

———— 1909a, Recent developments in southern Seward Peninsula: U.S. Geol. Survey Bull. 379, p. 267-301.

———— 1909b, The Iron Creek region: U.S. Geol. Survey Bull. 379, p. 302-354.

———— 1910, Geology and mineral resources of the Solomon and Casadepaga quadrangles, Seward Peninsula, Alaska: U.S. Geol. Survey Bull. 433, 234 p.

———— 1913, The Noatak-Kobuk region, Alaska: U.S. Geol. Survey Bull. 536, 160 p.

———— 1917, The Lake Clark-central Kuskokwim region, Alaska: U.S. Geol. Survey Bull. 655, 162 p.

———— 1926, Mineral industry of Alaska in 1924: U.S. Geol. Survey Bull. 783, p. 1–30.

———— 1929, Mineral industry of Alaska in 1926: U.S. Geol. Survey Bull. 797, p. 1–50.

———— 1930a, Mineral industry of Alaska in 1927: U.S. Geol. Survey Bull. 810, p. 1–64.

———— 1930b, Mineral industry of Alaska in 1928: U.S. Geol. Survey Bull. 813, p. 1–72.

———— 1932, Mineral industry of Alaska in 1929: U.S. Geol. Survey Bull. 824, p. 1–81.

———— 1933a, Mineral industry of Alaska in 1930: U.S. Geol. Survey Bull. 836, p. 1–83.

———— 1933b, Mineral industry of Alaska in 1931: U.S. Geol. Survey Bull. 844–A, p. 1–82.

———— 1933c, Past placer-gold production from Alaska: U.S. Geol. Survey Bull. 857–B, p. 93–98.

———— 1934a, Mineral industry of Alaska in 1932: U.S. Geol. Survey Bull. 857–A, p. 1–91.

———— 1934b, Mineral industry of Alaska in 1933: U.S. Geol. Survey Bull. 864–A, p. 1–94.

———— 1936, Mineral industry of Alaska in 1934: U.S. Geol. Survey Bull. 868–A, p. 1–91.

———— 1937, Mineral industry of Alaska in 1935: U.S. Geol. Survey Bull. 880–A, p. 1–95.

———— 1938, Mineral industry of Alaska in 1936: U.S. Geol. Survey Bull. 897–A, p. 1–107.

———— 1939a, Mineral industry of Alaska in 1937: U.S. Geol. Survey Bull. 910–A, p. 1–113.

———— 1939b, Mineral industry of Alaska in 1938: U.S. Geol. Survey Bull. 917–A, p. 1–113.

———— 1941a, Fineness of gold from Alaska placers: U.S. Geol. Survey Bull. 910–C, p. 147–272.

———— 1941b, Mineral industry of Alaska in 1939: U.S. Geol. Survey Bull. 926–A, p. 1–106.

———— 1942a, Occurrences of molybdenum minerals in Alaska: U.S. Geol. Survey Bull. 926–C, p. 161–207.

———— 1942b, Mineral industry of Alaska in 1940: U.S. Geol. Survey Bull. 933–A, p. 1–102.

Smith, P. S., and Eakin, H. M., 1911, A geologic reconnaissance in southeastern Seward Peninsula and the Norton Bay-Nulato region, Alaska: U.S. Geol. Survey Bull. 449, 146 p.

Smith, P. S., and Mertie, J. B., Jr., 1930, Geology and mineral resources of northwestern Alaska: U.S. Geol. Survey Bull. 815, 351 p.

Smith, S. S., 1917, The mining industry in the Territory of Alaska during the calendar year 1916: U.S. Bur. Mines Bull. 153, 89 p.

Smith, T. E., 1970, Gold resource potential of the Denali bench gravels, Alaska: U.S. Geol. Survey open-file rept., 19 p.

Smith, W. R., 1925, The Cold Bay-Katmai district: U.S. Geol. Survey Bull. 773, p. 183–207.

Spencer, A. C., 1906, The Juneau gold belt, Alaska: U.S. Geol. Survey Bull. 287, p. 1–137.

Steidtmann, Edward, and Cathcart, S. H., 1922, Geology of the York tin deposits, Alaska: U.S. Geol. Survey Bull. 733, 130 p.

Tailleur, I. L., 1969, Speculations on North Slope geology: Oil and Gas Jour., v. 67, no. 38, p. 215–220, 225–226.

Tarr, R. S., and Butler, B. S., 1909, The Yakutat Bay region, Alaska: U.S. Geol. Survey Prof. Paper 64, 183 p.

Thomas, B. I., 1957, Tin-bearing placer deposits near Tofty, Hot Springs district, central Alaska: U.S. Bur. Mines Rept. Inv. 5373, 56 p.

——— 1965, Reconnaissance sampling of the Avnet manganese prospect, Tanana quadrangle, central Alaska: U.S. Bur. Mines open-file rept., 8 p.

Thomas, B. I., and Berryhill, R. V., 1962, Reconnaissance studies of Alaskan beach sands, eastern Gulf of Alaska: U.S. Bur. Mines Rept. Inv. 5986, 40 p.

Thomas, B. I., and Wright, W. S., 1948a, Investigation of the Morelock Creek tin placer deposits, Fort Gibbon district, Alaska: U.S. Bur. Mines Rept. Inv. 4322, 8 p.

——— 1948b, Investigation of the Tozimoran Creek tin placer deposits, Fort Gibbon district, Alaska: U.S. Bur. Mines Rept. Inv. 4323, 11 p.

Thorne, R. L., Muir, N. M., Erickson, A. W., Thomas, B. I., Heide, H. E., and Wright, W. S., 1948, Tungsten deposits in Alaska: U.S. Bur. Mines Rept. Inv. 4174, 22 p.

Tuck, Ralph, 1933, The Moose Pass-Hope district, Kenai Peninsula, Alaska: U.S. Geol. Survey Bull. 849–I, p. 469–530.

——— 1938, The Valdez Creek mining district, Alaska, in 1936: U.S. Geol. Survey Bull. 897–B, p. 109–131.

U.S. Geological Survey, 1964, Mineral and water resources of Alaska: U.S. 88th Cong., 2d sess., Senate Comm. Interior and Insular Affairs, Comm. Print, 179 p.

——— 1965, Geological Survey research, 1965: U.S. Geol. Survey Prof. Paper 525–A, A1–A376.

Wahrhaftig, Clyde, 1958, Quaternary geology of the Nenana River valley and adjacent parts of the Alaska Range: U.S. Geol. Survey Prof. Paper 293–A, p. 1–68.

——— 1965, Physiographic divisions of Alaska: U.S. Geol. Survey Prof. Paper 482, 52 p.

——— 1968, Schists of the central Alaska Range: U.S. Geol. Survey Bull. 1254–E, p. E1–E22.

——— 1970a, Geologic map of the Healy D–2 quadrangle, Alaska: U.S. Geol. Survey Geol. Quad. Map GQ–804, scale 1:63,630.

——— 1970b, Geologic map of the Healy D–3 quadrangle, Alaska: U.S. Geol. Survey Geol. Quad. Map GQ–805, scale 1:63,360.

——— 1970c, Geologic map of the Healy D–4 quadrangle, Alaska: U.S. Geol. Survey Geol. Quad. Map GQ–806, scale 1:63,360.

——— 1970d, Geologic map of the Healy D–5 quadrangle, Alaska: U.S. Geol. Survey Geol. Quad. Map GQ–807, scale 1:63,360.

———— 1970e, Geologic map of the Fairbanks A-2 quadrangle, Alaska: U.S. Geol. Survey Geol. Quad. Map GQ-808, scale 1:63,360.

———— 1970f, Geologic map of the Fairbanks A-3 quadrangle, Alaska: U.S. Geol. Survey Geol. Quad. Map GQ-809, scale 1:63,360.

———— 1970g, Geologic map of the Fairbanks A-4 quadrangle, Alaska: U.S. Geol. Survey Geol. Quad. Map GQ-810, scale 1:63,360.

———— 1970h, Geologic map of the Fairbanks A-5 quadrangle, Alaska: U.S. Geol. Survey Geol. Quad. Map GQ-811, scale 1:63,360.

Waters, A. E., Jr., 1934, Placer concentrates of the Rampart and Hot Springs districts: U.S. Geol. Survey Bull. 844-D, p. 227-246.

Wayland, R. G., 1961, Tofty tin belt, Manley Hot Springs district, Alaska: U.S. Geol. Survey Bull. 1058-I, p. 363-414.

Wedow, Helmuth, Jr., Killeen, P. L., and others, 1954, Reconnaissance for radioactive deposits in eastern interior Alaska, 1946: U.S. Geol. Survey Circ. 331, 36 p.

Wedow, Helmuth, Jr., and others, 1953, Preliminary summary of reconnaissance for uranium and thorium in Alaska, 1952: U.S. Geol. Survey Circ. 248, 15 p.

Wedow, Helmuth, Jr., White, M. G., and others, 1954, Reconnaissance for radioactive deposits in east-central Alaska, 1949: U.S. Geol. Survey Circ. 335, 22 p.

Wells, F. G., 1933, Lode deposits of Eureka and vicinity, Kantishna district, Alaska: U.S. Geol. Survey Bull. 849-F, p. 335-379.

West, W. S., 1953, Reconnaissance for radioactive deposits in the Darby Mountains, Seward Peninsula, Alaska, 1948: U.S. Geol. Survey Circ. 300, 7 p.

West, W. S., and Benson, P. D., 1955, Investigations for radioactive deposits in southeastern Alaska: U.S. Geol. Survey Bull. 1024-B, p. 25-57.

West, W. S., and Matzko, J. J., 1953, Buckland-Kiwalik district, 1947, in Gault, H. R., and others, 1953, Reconnaissance for radioactive deposits in the northeastern part of the Seward Peninsula, Alaska, 1945-57 and 1951: U.S. Geol. Survey Circ. 250, p. 21-27.

White, D. E., 1942, Antimony deposits of the Stampede Creek area, Kantishna district, Alaska: U.S. Geol. Survey Bull. 936-N, p. 331-348.

White, M. G., 1952a, Reconnaissance for radioactive deposits along the upper Porcupine and lower Coleen Rivers, northeastern Alaska: U.S. Geol. Survey Circ. 185, 13 p.

———— 1952b, Radioactivity of selected rocks and placer concentrates from northeastern Alaska: U.S. Geol. Survey Circ. 195, 12 p.

White, M. G., and Killeen, P. L., 1953, Reconnaissance for radioactive deposits in the lower Yukon-Kuskokwim highlands region, Alaska, 1947: U.S. Geol. Survey Circ. 255, 18 p.

White, M. G., Nelson, A. E., and Matzko, J. J., 1963, Radiometric investigations along the Taylor Highway and part of the Tanana River, Alaska: U.S. Geol. Survey Bull. 1155, p. 77-82.

White, M. G., and Stevens, J. M., 1953, Reconnaissance for radioactive deposits in the Ruby-Poorman and Nixon Fork districts, west-central Alaska, 1949: U.S. Geol. Survey Circ. 279, 19 p.

White, M. G., West, W. S., and Matzko, J. J., 1953, Reconnaissance for radioactive deposits in the vicinity of Teller and Cape Nome, Seward Peninsula, Alaska, 1946-47: U.S. Geol. Survey Circ. 244, 8 p.

Williams, J. A., 1951, Memorandum report on a radiometric investigation of the Connell property, 28 mile, Yukon River, Circle district: Alaska Dept. Mines unpub. rept. PE 51–1, 5 p.

Williams, J. R., 1962, Geologic reconnaissance of the Yukon Flats district, Alaska: U.S. Geol. Survey Bull. 1111–H, p. 289–331.

Wright, C. W., 1904a, The Porcupine placer mining district: U.S. Geol. Survey Bull. 225, p. 60–63.

———— 1904b, The Porcupine placer district, Alaska: U.S. Geol. Survey Bull. 236, 35 p.

Wright, F. E., and Wright, C. W., 1906, Lode mining in southeastern Alaska, 1907: U.S. Geol. Survey Bull. 284, p. 30–54.

INDEX OF LOCALITIES

201

* U.S. GOVERNMENT PRINTING OFFICE: 1973—515—659/72

ALASKA PENINSULA REGION
ALEUTIAN ISLANDS REGION
BERING SEA REGION
BRISTOL BAY REGION
COOK INLET – SUSITNA REGION
1 Anchorage district
2 Redoubt district
3 Valdez Creek district
4 Willow Creek district
5 Yentna district
COPPER RIVER REGION
6 Chistochina district
7 Nelchina district
8 Nizina district
9 Prince William Sound district
10 Yakataga district
KENAI PENINSULA REGION
11 Homer district
12 Hope district
13 Seward district
KODIAK REGION
KUSKOKWIM RIVER REGION
14 Aniak district
15 Bethel district
16 Goodnews Bay district
17 McGrath district
NORTHERN ALASKA REGION
18 Barrow district
19 Canning district
20 Colville district
21 Lisburne district
22 Wainwright district
NORTHWESTERN ALASKA
23 Kana district
24 Noatak district
25 Selawik district
26 Shungnak district

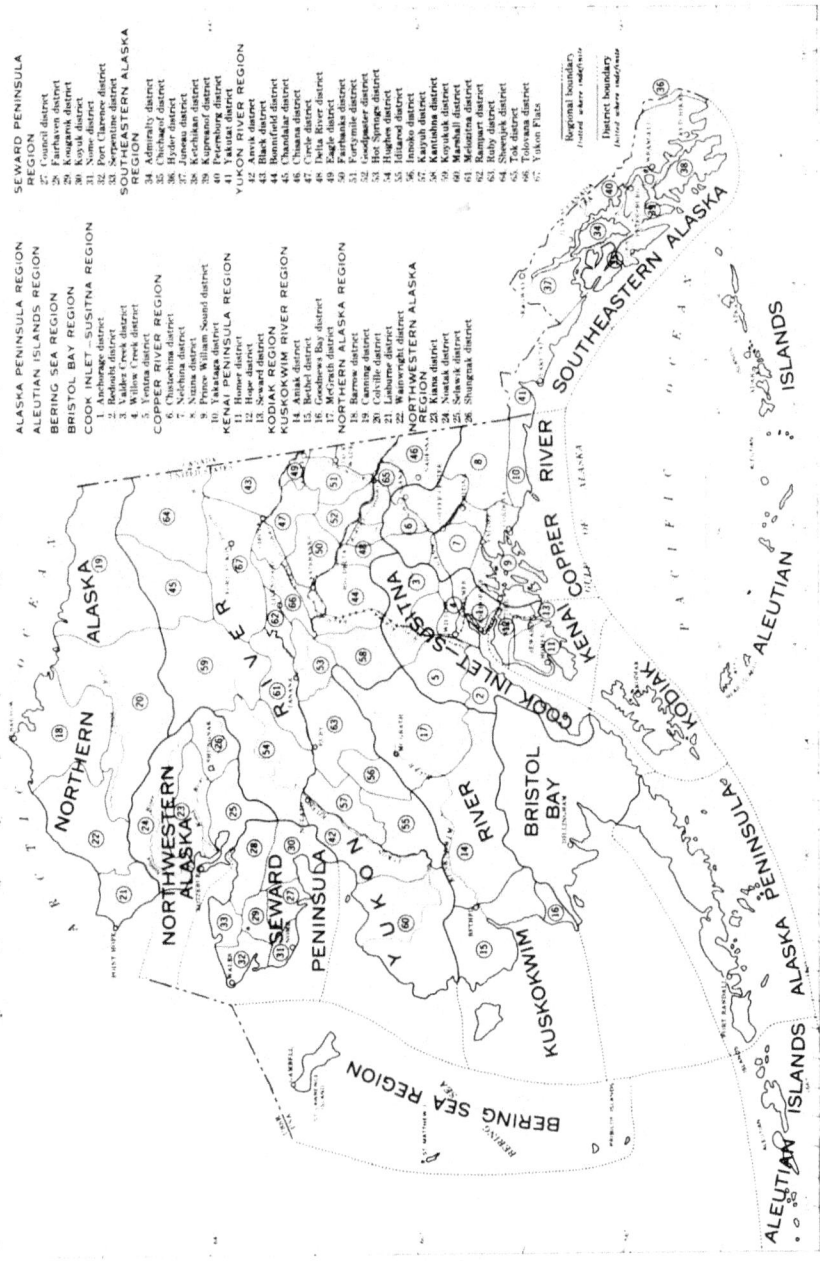

SEWARD PENINSULA
REGION
27 Council district
28 Fairhaven district
29 Kougarok district
30 Koyuk district
31 Nome district
32 Port Clarence district
33 Serpentine district
SOUTHEASTERN ALASKA
REGION
34 Admiralty district
35 Chichagof district
36 Hyder district
37 Juneau district
38 Ketchikan district
39 Kupreanof district
40 Petersburg district
41 Yakutat district
YUKON RIVER REGION
42 Amik district
43 Black district
44 Bonnifield district
45 Chandalar district
46 Chisana district
47 Circle district
48 Delta River district
49 Eagle district
50 Fairbanks district
51 Fortymile district
52 Goodpaster district
53 Hot Springs district
54 Hughes district
55 Idttarod district
56 Innoko district
57 Kantishna district
58 Kasyuh district
59 Koyukuk district
60 Marshall district
61 Melozitna district
62 Rampart district
63 Ruby district
64 Shungnak district
65 Tok district
66 Tolovana district
67 Yukon Flats

Regional boundary
Dotted where indefinite

District boundary
Dotted where indefinite

From Berg and Cobb (1967)

MINING REGIONS AND DISTRICTS

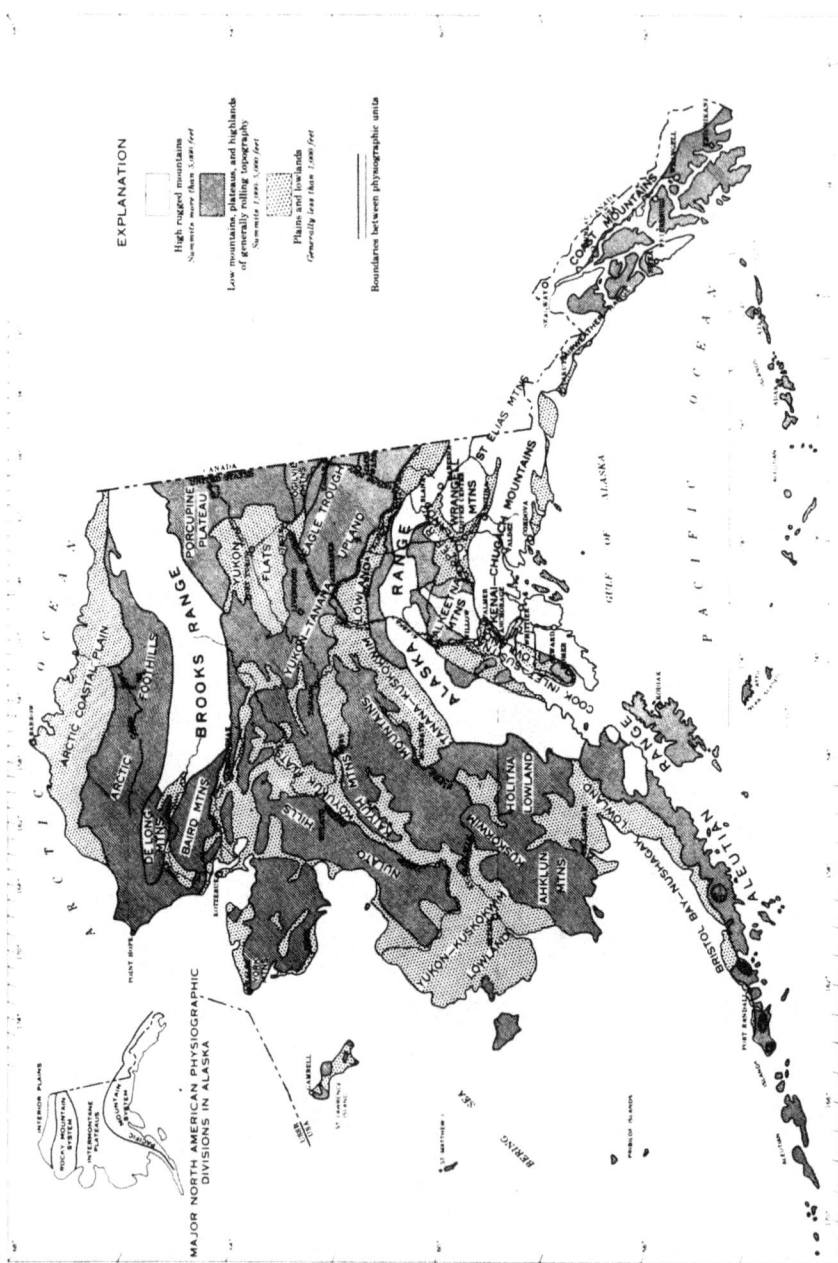

PHYSIOGRAPHIC PROVINCES

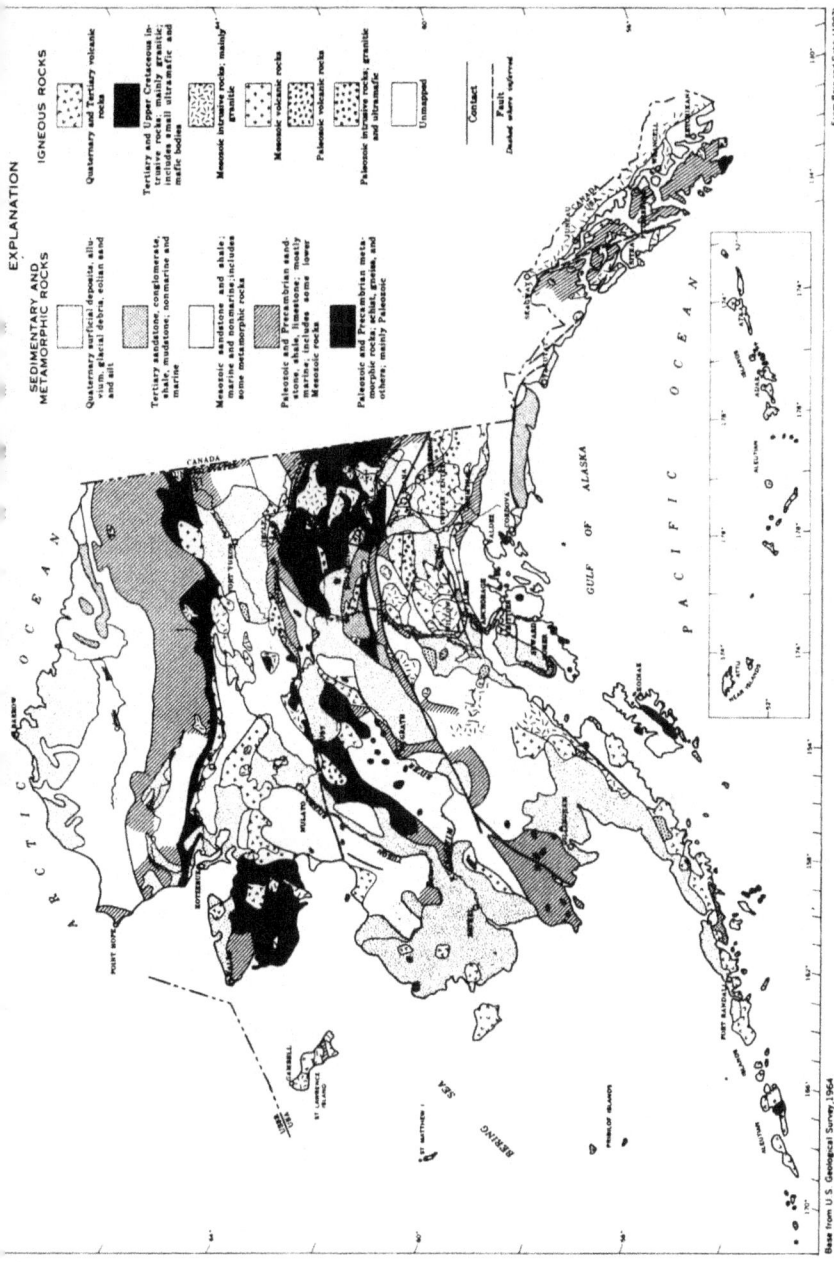

GENERALIZED GEOLOGIC MAP

MAPS OF ALASKA SHOWING MINING REGIONS AND DISTRICTS, PHYSIOGRAPHIC PROVINCES, AND GEOLOGY

Base from U.S. Geological Survey, 1954

From Berg and Cobb (1967)

www.ingramcontent.com/pod-product-compliance
Lightning Source LLC
Chambersburg PA
CBHW060549200326
41521CB00007B/541